Modern Silver

MODERN

Crown Publishers, Inc., New York

SILVER

THROUGHOUT THE WORLD 1880-1967

Graham Hughes

FRONT ENDPAPER
Shows candelabra (*left*) in silver-plated brass. **DM** Gertraud von
Schnellenbühl after 1911, ht 19 in. **O** Stadtmuseum Munich;
(*right*) in sterling silver **DM** David Mellor, Sheffield 1964,
diam. 3 ft 6 in. **O** The Worshipful Company of Goldsmiths,
London.

BACK ENDPAPER
(*Left*) a textured sterling silver jug given by Mrs Winifred
Armstrong to Fitzwilliam College, Cambridge, to celebrate its
achieving full status. **DM** Gerald Benney, London 1965; and
(*right*) a large cup and cover in silver and translucent enamel.
DM Philippe Wolfers 1904 **O** Ghent Museum.
On the title page is the 'Helicopter World' Trophy in aluminium.
DM Geoffrey Clarke, Stowmarket 1966.

JACKET
A detail of the large bowl number 33 **DM** Bolin Stockholm 1916.

© Graham Hughes 1967
First published in 1967 in the United States of America by
Crown Publishers, Inc., 419 Park Avenue South, New York,
N.Y. 10016, and in Great Britain by Studio Vista Limited,
Blue Star House, Highgate Hill, London N19
Library of Congress Catalog Card Number 67–26050
Printed and bound in Great Britain by W. S. Cowell Ltd

Contents

Introduction   page 7
Geography   9
Training and technique   55
Genesis and Exodus   67
Stainless steel   87
Pioneers   93
Patrons and design   129
Biographies   219
Acknowledgements   254
Bibliography   255
Index   256

Introduction

In 1891 the great Belgian artist Henri van de Velde was in Brussels at the opening banquet of the eighth Salon des XX. There he heard of the writing of the British idealists, William Morris and John Ruskin, and he first saw the works of Walter Crane. He began to feel the importance of applied art, abandoned painting, his first love, and taught himself to be an architect. This visionary in fine art had accepted the discipline of utility, and his life both before and after his conversion stands as a good symbol for this book.

Till the 1890s many leading theorists thought it nobler to copy than to create: look back to the great periods, so the argument ran, and one can't go far wrong. Even William Morris, who wanted original design, deliberately copied medieval production methods for his firm, Morris and Company, and used designs by the Pre-Raphaelites, whose declared aim was to revive the old Italian ideas. But with the *art nouveau* movement these inhibitions went to the winds. It was originality, not imitation that counted, and a new quality of fun and enjoyment replaced the solemn piety of the early visionaries. As the twentieth century progressed, design became increasingly personal until it was eventually disciplined by new techniques of mass-production and new ideas about financial economy. Today in silver a happy balance prevails: an expensive material now requires designs that are not merely commonplace, whilst practical economy brings sense to the exaggerations which were still possible forty or sixty years ago.

In 1907 Van de Velde wrote, 'As for me, I still prefer the hand-woven table-cloth to the machine-made product, hand-made silver to machine-stamped cutlery.' Continental *art nouveau*, to which he contributed so much, looked forward artistically, but backward technically. Its leaders soon realized this and mostly abandoned the style: from 1921 Van de Velde spent his time perfecting one of Europe's most functional buildings, the Kröller-Müller Museum, which has no fantastic frills, and no hint that its designer had no professional training. During his long life (1863–1957) he had created some of the wildest fantasies but he finally taught himself an impressive professional austerity. He worked in Germany, Norway, Switzerland, France and Holland, as well as Belgium, and he can be called truly international. He was one of many brilliant people whose style followed similar courses, and who eventually achieved the balance between form and function that we now call modern design.

I have tried to pin down the influence of important figures like Van de Velde on the production of small trifles for the home, a story which has often been told before in general terms. And I have investigated methods of production, the swift and predictable rise of machining, and the determined and quite unexpected resurgence today of creative handwork, the result of a healthy reaction by the consumer against the spread of uniformity.

I have emphasized the *art nouveau* movement whose virtuoso designs fascinate us today, though their social impact was then extremely narrow; the next wave of fashion was an undigested functionalism, an obsession with the machine which, in retrospect, seems both harsh and insensitive, so I have selected from the period between the wars only the strongest examples to suggest the battle between man and mechanization. Since 1945 peace has been achieved: we have come to accept that a man can work with a machine just as well as he can with his own hands – and we no longer yearn for the magic of handwork in the way that William Morris used to do.

Only when the machine and the mechanical process dominate their organizer does he become a mere operative instead of a creative craftsman. Wherever the design seems right I have included cheap, mass-produced work, thus covering factory efficiency as well as the richness of handwork. I have shown stainless steel as well as gold and silver and electroplate because it is nearly always the same people who handle them. But I have omitted clocks and watches because they come from different factories and countries. I have tried to represent the main types of modern post-war production, which has so often been used to bring colour and glamour to our drab, everyday lives, and I have particularly tried to be enterprising and adventurous in my choice.

I have concentrated on the impact of large photographs of good pieces, which I hope will be memorable for posterity, rather than on giving an indigestible catalogue of tiny pictures, too small to convey the character of silver. I have attacked both the visual story and the narrative from several angles, personal, artistic, chronological and technical, with the aim of giving a representative account of this century's silver. To trace the story of one designer in detail, and omit another entirely, is unfair but inevitable. Any book about living people is bound to be invidious, and I hope all my friends who find themselves entirely left out will forgive me and realize that the best book is not the one that contains everything, but the one which is selective, stimulating, and therefore easy to read.

I am particularly grateful to friends overseas who have helped me to find photographs: Sven Boltenstern, jeweler of Vienna; Max Fröhlich, silversmith in Zürich; Sven Arne Gillgren, silversmith in Stockholm; Olof Gummerus of Helsinki, who has done so much to achieve world fame and eminence for modern Finnish design, and Dr Kobell of the remarkable Landesgewerbeamt museum in Stuttgart, a unique collection of unpretentious private objects, typical not of the aesthete but of every man. Without the generous help of these and many others, the book could not have claimed to be comprehensive.

My colleagues at Goldsmiths' Hall, Susan Hare, Sue Anderson and Anne Colchester, have generously allowed the book to impinge on their already extensive efforts to help modern metalwork. Mrs Amanda Drybrough edited some of the biographical section with her usual flair. Many designers and firms have kindly given me their histories, so that much of the information here has never before appeared in print. Lastly, I thank the photographers to whom I have made acknowledgement separately. Silver is perhaps the most difficult of all photographic subjects; the same piece of silver under different lights may look like chromium plate or putty, but I hope my reproductions give some of the unique beauty and lustre of this noble metal. Silver is expensive, and the people who design and make it therefore take particular trouble; I hope this book will convince its readers that modern silver is a creative art.

Afriston, 1967, Graham Hughes

Geography

A survey of some of the world's most influential firms

*Silver objects, like those formed of clay or glass, should
perfectly serve the end for which they have been formed.*

Christopher Dresser, designer (*Principles of Design* 1871/2).

*The Church has been one of the great spending powers of
our time. It must have spent millions of money turning
authentic into Bunkum.*

W. R. Lethaby (1857–1931). Professor of Design, Royal College of Art; first principal London
County Council, Central School of Arts and Crafts, surveyor Westminster Abbey. *Collected
Writings*.

JENSEN, Denmark

Although the world is getting smaller every day, silver in each producing country still has its
own character. One leading firm in each of the main producing centres serves as a good introduction to a complete survey – the hors-d'œuvre, whetting the appetite by showing what is
different in each place, before the main course which shows what modern silver has in common
everywhere. But with at least one firm, this is more than the hors-d'œuvre – if Georg Jensen
had never lived, Scandinavian modern silver might never have brought the ideals of William
Morris to practical realization, and silversmiths today might still be concentrating, as they
were a century ago, on copies, not creation.

If Jensens represent quality, Bolins show the survival of aristocracy; if Pott is determined
refinement and specialization, WMF is planned diversity. If Tiffany's are eccentricity tamed,
the Swedish GAB group shows sanity so great as to verge on distinction.

In most countries the old British dream survived of enlightened hand versus evil machine.
It had grown up with the development of machinery in England around 1760 and it first began
to be more nightmare than dream in Nottingham in the 1810s when the poor simple factory
workers – the 'Luddites' – thought they could keep their jobs open by smashing the machines
which seemed to be starving them. Jensen showed the way out: men can master machines and
the hand can outwit the tool.

George Jensen was born on 31 August 1866. His name is still famous not as a great man, not
even as a great innovator, but because he succeeded where everyone else had failed. Fine
modern silver was not his idea – it first appeared, in fact, in England; nor did he invent a new
style – *art nouveau* too had begun in England. What he did achieve, and it proved impossible
for other people more brilliant than him, was success. He established modern design commercially. He revived handwork in the machine age.

There were virtuoso designers contemporary with Jensen who rightly saw that rich people
would buy their personal handwork in preference to run of the mill mediocrity – Charles
Rennie Mackintosh of Glasgow who started the whole modern movement in design and architecture about 1880; René Lalique of Paris whose jewels and glass captivated the *belle monde*
including Sarah Bernhardt; Victor Prouvé and Emile Gallé of Nancy whose furniture was the
most uncomfortable and glass the most fragile of its day; Philippe Wolfers of Brussels, the
crown jeweler turned sculptor turned interior designer; Louis Comfort Tiffany of New York,
latterly Art Director of his father's firm and generous host to his artists' colony at Oyster Bay,
Long Island, inventor of the pearly streaky glass he called 'favrile'; Gaudì of Barcelona, even
Sant' Elia of Milan, the visionary architect whose superb ideas never got built – all these
pioneers of visual freedom at the turn of the century were bigger men than Jensen, more inspired, more original and better known. But these artists simply wanted to create for their
élite customers, and their social purpose was secondary to their urge for personal expression.
They were wonderful freaks, so exotic that their work died with them.

'He who can, does; he who cannot, teaches.' George Bernard Shaw's advice is relevant. Beside the great inventors were the great missionaries. About 1880, many aesthetes, frightened of the machine, tried to save western culture by preaching. William Morris of England, one of the first and most influential, started the arts and crafts movement to rouse public opinion. His own Morris and Company was quite successful commercially, lasting as late as 1940. C. R. Ashbee's Guild of Handicrafts in London and in Chipping Campden was more typical, well meaning but unrealistic – finished in 1907. The Artificers Guild of the silversmith Edward Spencer was another – finished in 1936. The Omega Workshops of the distinguished English critic Roger Fry harnessed some of the best London painters and sculptors to make some of the worst furnishings ever. Result – finished in 1919. In the other great European empire, Austria-Hungary, Vienna's Wiener Werkstätte, founded in 1903, employed before the 1914 war as many as sixty craftsmen and powerfully nourished the local modern style called Sezession. But they too were bankrupt in 1932. In Germany the modern style was called Jugendstil; and the various regional societies like the Deutsche Werkstätte of 1898, or the Deutscher Werkbund of 1907 culminated in the Bauhaus of Weimar, Dessau and Darmstadt, dedicated to efficient functional design, killed by Hitler in 1933. Only the Bauhaus was entirely realistic and it liked machines, not hands: Naum Slutzky, the resident Bauhaus jeweler who died in England in 1965, always felt the pressure to mass-produce jewels though he knew the best jewels must be hand-made, one by one.

All these educational fantasies were badly hit by the 1914 war, and the 1929 slump, but even so they seem in retrospect to have been too sentimental: handcrafts are of no importance unless they are properly organized, supported with a convincing balance sheet; and the best organizers of the time were running factories, not trying to stop them.

Between the lone creators who didn't make a movement and the brave associations who didn't make a penny was Jensen. Impulsive, childlike and incessantly creative, he opened his tiny shop at 36 Bredgade, Copenhagen in 1904. Only twenty-five years later the man had become a business with a staff of 250. He had already achieved his great ambition, to make handsome modern silver popular everywhere.

His liberal patrons made little fuss, but actually hit the bull's-eye called progress. If Scandinavian design might not have happened without Jensen, Jensen might not have been noticed except in Scandinavia: and he might not have survived but for the enlightened Hostrup-Pedersen family whose informal backing finally, in 1922, became ownership of the whole Jensen enterprise.

On the one hand, the Jensen firm is a remarkable combination of family piety devoted to the founder's ideals, still supported by his son and others of his descendants; and on the other, of wise management by the Hostrup-Pedersen family. Their father was an engineer, designing power stations and electrical installations throughout Denmark, and he evidently found the silversmithy a rewarding, though expensive, antidote to his own heavy industry. His three sons, two of whom manage the Copenhagen factory and one the London shop, likewise were none of them trained in silver or design, and therefore bring to the firm a splendid breadth of vision coupled with a less than passionate respect for the profit motive. Profits there are, and they are increasing, but the firm believes it has something more important than money to offer the world, and the world is slowly coming to agree.

The revival of Denmark's self respect was partly due to Bishop Grundtvig in the 1860s, whose reforms created a strong economy. After the war in 1864, when Denmark lost a fifth of her land to Germany, she was broken. Bishop Grundtvig in his enlightened Christianity started the folk high school movement, thus bringing knowledge and culture to the masses. These schools soon started technical farming. A spiritual revival came with commercial stability: by the late nineties the Danes were able to buy art and luxury. At the same time, the future British King Edward VII married the Danish Princess Alexandra, no doubt helping to make Denmark aware of the British arts and crafts movement which became so important an inspiration in Scandinavia.

By the turn of the century, a heavy bulbous Danish *art nouveau* style had been developed by such designers as Thorvald Bindesbøll, the architect who specialized in ceramic building decoration and large pots and who designed the labels still in use on the Carlsberg beer bottles;

or Knud Engelhardt who designed the Copenhagen trams, or Arnold Krog of the Royal Copenhagen Porcelain Company. These three all dabbled in silver, and of course there were excellent old silver firms with conservative products like Michelsen and Dragsted of Copenhagen, Hingelberg of Aarhus, Hans Hansen of Kolding, and Cohr of Fredericia. But it was the new man Jensen who determined to make only modern work and whose modern production was, therefore, on a huge scale compared with the others. Even his influential first master, Mogens Ballin (who later abandoned art in favour of religion), with his small traditional workshop, had not in any way fired the national imagination. Jensen with his early colleague the painter Johan Rohde, who also died in 1935, transformed the picture. In 1935 at the Brussels World Fair, the two official Danish guests were Nils Bohr the physicist, and Georg Jensen; it was pleasing that new science had been coupled with an old craft, and it was a fitting honour to Jensen.

Jensen's personal story belongs in a later chapter with those other few designers who have managed to develop an unmistakable personal style, and therefore left their mark on the century's silversmithing. Rohde never became a director of the firm, and one may imagine that Jensen moved slowly into the managerial field, leaving designs more and more to Rohde and his successors. Rohde's most famous design was probably the plain pitcher of 1920, first exhibited in the Paris 1925 Salon d'Autonne, and it is tempting to credit Rohde with simple dignity, in contrast to Jensen's own incessant grapes and nuts. But the evidence of the very large number of drawings preserved in the Jensen factory in Copenhagen is otherwise; the two men were interdependent, both capable of very elaborate designs as well as of simplicity. Only a practised eye can detect the difference between the two designers, and such an eye anyway probably knows by heart the few hundred production patterns.

The sister of Jensen's third wife married Gundorph Albertus, the first of the many distinguished artists upon whom the continuity of the firm depended. He joined Jensens as early as 1911, having studied sculpture at the Copenhagen academy of fine arts, and he became the director of production. Arno Malinowski is a sculptor too, but his training as an engraver shows particularly in the precision of the die-stamped animal brooches for which the firm became famous in the 1930s. More important than either as a lasting influence is Harald Nielsen, who joined Jensen in 1909, and is still in the closest touch with the firm's direction. His work is quite distinctive – more functional, more related to the contemporary Bauhaus ideas of geometrical precision; it was typical of the best Jensen production between the wars. A distinguished foretaste of the future was Sigvard Bernadotte, son of the King of Sweden. He joined Jensens in 1931 and designed mainly for them until 1947 when he began his industrial design; the first of the Jensen team whose style can be called hard rather than soft, his pieces have a frigid elegance, the signature for which used to be his austere ornamentation of incised parallel lines. If Nielsen brought Jensens into the 1920s, Bernadotte imposed the international anonymity of the 1930s. These were the years of his greatest influence on the firm. But he still designs for them even after being president of the Swedish Society of Industrial Designers, and of the international society. In 1967 he moved further from silver, becoming a director of the London design consultancy, Allied Industrial Designers.

During the German occupation of Denmark silver was unobtainable, and Jensen built a new factory in the suburbs to produce stainless steel, a sharp break with their past, and a fortunate chance result of war-time disaster. After the war, the number of factory production employees seemed about constant at 250, and the firm's reputation for impeccable craftsmanship was interpreted by some as 'never corrupted by human hand' and by others as 'entirely hand-made'. Both were true, and still are: some products, particularly stainless steel, are suited to and indeed improved by very highly organized mass-production; others, including most of the original Jensen designs with their beautiful soft hammer-marked surface, have to go slowly through the workshop receiving individual love and care one at a time. Jensens had their retail shops in several countries, sometimes independent of the parent, sometimes wholly owned, sometimes run by relations of the family; with increasing competition both at home and abroad, and with a huge range of designs in production, very few of which measured up to modern fashion but most of which were still good sellers, great policy decisions were made. Jensens would be organic not static, and the Jensen style would come to mean not just the designs of the old man himself, but an increasing variety of the best that could be done in gold, silver and steel.

Magnus Stephensen, architect of many fine buildings, designed some of the award winning utility goods of the 1950s – saucepans, boxes, serving dishes; he, like many architects, flourishes on stainless steel, where precision is everything, and if straight lines are to be endured, Stephensen's finish gives them some allure. But the first of the Jensen team since Jensen himself and Rohde to show a completely personal inspiration, more typical of southern Mediterranean abandon than of northern restraint, is the young sculptor Henning Koppel.

He joined Jensens in 1945 and first became famous ten years ago. At that time, top aesthetes were still admiring the ultimate development of Bauhaus theory, that whatever was easiest to produce and most practical to use, automatically looked best, and if one happened oneself to think that it was a little boring, that showed one's own vulgar taste, not any lack of inspiration in the object itself. Koppel is not a revolutionary as painter and sculptor – his works in that field seem almost tame in the context of the wave of fashionable tachism, action painting, pop art and op art that have successfully assaulted the myriad private galleries of London and New York; but as silversmith he represented something quite new in Scandinavia, just as Jensen himself had done fifty years before. He again made personal creation respectable, and the noble curves of his great free form bowls and pots have brought dignity as well as publicity to the firm.

To list all the Jensen prizes would be ridiculous, but Koppel stands as a fair representative of the high standing won by Jensens in the art world: in 1947 he won 3rd prize in the bank note competition; in 1953 the National Art Gallery Drawing-prize and Scandinavia's chief design award – the Lunning Prize; in 1951, 1954 and 1957 the highest Italian award, a gold medal at the Milan Triennale; in 1963 the US International Design Award and 1st prize in the Danish postage stamp competition (his idea was not eventually used) and the Golden Spoon prize in Munich; and in 1964 3rd prize for a chair in the Danish Beech Wood competition and 2nd prize for dice in the Danish tourist industry souvenir competition. In 1966 Jensens daringly, for the first time, entered the New York Diamonds-International contest, and Koppel's extraordinary jewels won three of the twenty-one awards, themselves chosen from 1400 entries from twenty-three different countries.

Koppel works full time for Jensens, which means that he spends as much time on them as his instincts require, and has permanent access to their factory studio. He is not expected to work for other silversmiths, and is much more valuable to Jensens than an occasional consultant designer; but he has tried his hand at many things including lamps and clocks, and since 1961 he has designed for Bing & Grøndahl Porcelain. Jensens positively prefer their designers to fertilize their talent by travelling and working on all sorts of varied projects. When Koppel had a one-man show in 1966 at the Jensen New York shop, the firm, of course, sent him there; and he came to the Jensen Centenary Exhibition at Goldsmiths' Hall London.

Jensen's own son, Søren Georg Jensen is unusual: most sculptors joining Jensen get sucked into the consuming vortex of silversmithing, either because they are better at it than they are at sculpture, or because they succumb to the charm of the Hostrup-Pedersens. Not so Søren. He remains a sculptor first, and his work is dominant in the splendid new art museum at Louisiana, twenty miles north of Copenhagen. But worldly wisdom as well as family piety have given him a special place in the firm where he is now the artistic manager. His silver production is small but each of his designs is memorable. The same could be said of the firm as a whole, in contrast for instance to one of the traditional factories in Providence, Rhode Island, or Birmingham, England, where profit not quality may be the dominant aim.

Nanna Ditzel has brought a new clarity and purpose to the firm's jewelry, and indeed to Scandinavian modern jewelry as a whole, which before her hardly existed. As well as the bright stars in the Jensen firmament, there are the dazzling comets, here today and gone tomorrow, who have won the firm's occasional competitions or undertaken some temporary commission: Ib Bluitgen, Tias Eckhoff, Bent Gabrielsen Pedersen, Erik Herløw, Tuk Fischer, Jørgen Dahlerup, Gert Holbek, and Ibe Dalquist.

Since Jensen's death, the firm has become much more international, much more highly mechanized and much more diverse. It has an influence out of all proportion to its modest size. It sells throughout the world and can honestly claim to be an international power-house for quality.

1 Coffee set 120 mm
  D Johan Rohde 1906
  M Georg Jensen, Copenhagen
  O National Museum, Stockholm

2 Acorn pattern
  D Rohde 1915
  M Jensen

3 Acanthus pattern
  D Rohde 1917
  M Jensen

4 **D** Rohde 1918
  M Jensen

4

5

6

7

8

5 Fish dish 760 mm
   D Rohde 1919
   M Jensen

6 Fish dish 580 mm
   D Harald Nielsen 1931
   M Jensen
   O Kunstindustrimuseum

7 **D** Rohde 1937
   M Jensen

8 Regency pattern
   D Nielsen 1947
   M Jensen

9 **D** Rohde 1927
   M Jensen

10 Pitcher 230 mm
   D Rohde 1920
   M Jensen
   O Kunstindustrimuseum,
   Copenhagen; National Museum,
   Stockholm; National Gallery,
   Melbourne

11 **D** Nielsen 1935
   M Jensen

9

10

11

12   **13**   **14**

15

16

12 **D** Nielsen 1927
   M Jensen

13 **D** Nielsen 1946
   M Jensen

14 **D** Bernadotte 1939
   M Jensen
   O National Gallery, Melbourne;
        Cranbrook Academy of Art,
        Michigan

15 **D** Sigvard Bernadotte 1932
   M Jensen

16

16 **D** Nielsen 1927
   M Jensen

17 **D** Nielsen 1935
   M Jensen

17

18

19

20

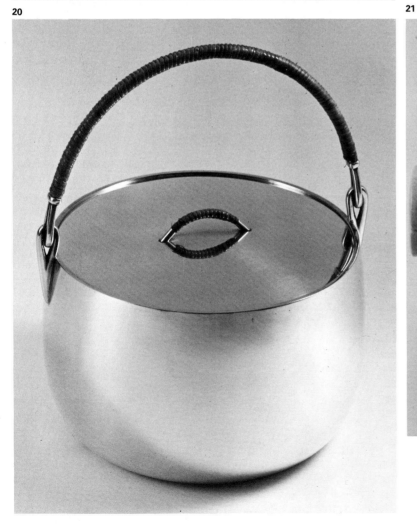

21

18 Salt and pepper shaker
  D Bernadotte 1943
  M Jensen

19 Condiment set, the shapes
   influenced by a chemical retort.
  D Søren Georg Jensen 1951
  M Jensen

20 Ice pail
  D Magnus Stephensen 1952
  M Jensen
  O Kunstgewerbemuseum, Zürich;
     Museum of Modern Art, New
     York; Kunstindustrimuseum,
     Oslo; Kunstindustrimuseum,
     Copenhagen.
     Grand Prix, Milan Triennale
     1954

21 Dish 240 mm
  D Stephensen 1962
  M Jensen

22 Teapot 135 mm
   D Henning Koppel 1952
   M Jensen

23 Pitcher
   D Koppel 1950
   M Jensen

24 Pitcher 287 mm
   D Koppel 1952
   M Jensen

22

23

24

19

25

26

25 Pitcher
   D Koppel 1956
   M Jensen

26 Fish platter
   D Koppel 1954
   M Jensen

27 Fish platter
   D Koppel 1956
   M Jensen
   O Worshipful Company of
        Goldsmiths, London

28 **D** Koppel 1950
   M Jensen Diam. 15½ in.
   O The Worshipful Company of
        Goldsmiths

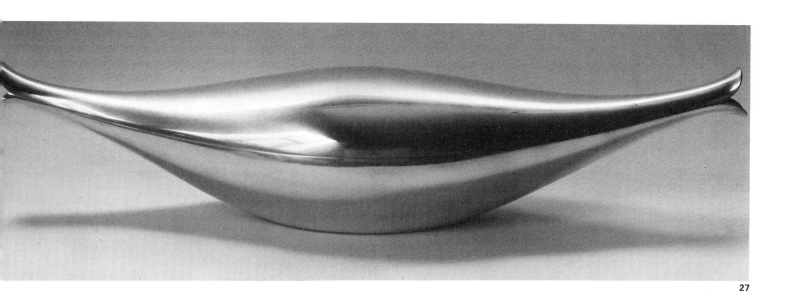

29 Tea and coffee set
   D Koppel 1963
   M Jensen

30 Caravel pattern
   D Koppel 1957
   M Jensen
   O Metropolitan Museum of Art,
       New York; LGA Museum,
       Stuttgart; Neue Sammlung,
       Munich.

31 Casserole
   D Koppel 1961
   M Jensen
   O LGA Museum, Stuttgart

BOLIN, Sweden

Still in the North is a very different character, Bolin, the old family firm who with Tillander, now of Helsinki, and Fabergé, now dead, shared the incredible riches of the Russian aristocracy; of the three, Bolin was once the most prominent, and it was actually they who introduced Fabergé to Imperial favour, little suspecting his subsequent fame. Firms in this most personal of trades depend particularly on the quality of their management – indeed, very few firms seem to survive a change of management, and the sort of management which succeeds seems to need the sort of devotion which only family spirit can give.

At the beginning of the eighteenth century, the Bolin family were shipowners on the west coast of Sweden, near Gothenburg. They moved to Stockholm at the end of the century and continued their shipping business there. John Bolin, the great-grandfather of the present head of W. A. Bolin had twelve children. In addition to his passion for the sea, he also enjoyed art, especially music: within his family there were no less than three quartets which at the time were much appreciated and well known in the social circles of Stockholm. In 1836 John Bolin left for Madeira on a voyage which was to be his last. He had taken with him his eldest son, and on their way home they ran into thick fog in the English Channel, colliding with a British ship and sinking. The whole crew was saved, except the two Bolins, father and son. Their remains were later recovered and buried at Hastings in England. The Bolin shipping company was then sold by John Bolin's widow who still lived in Stockholm, in a house which was eventually demolished in 1919.

Several of John Bolin's sons found their way to other countries. One served with the Swedish Foreign Office and was sent to the Swedish Embassy in London. Later he was asked to establish a Swedish Consulate at Malaga, Spain, an important port for Swedish shipping. He remained there as Swedish consul, married a French woman and had two sons. After a few years in Spain he died. His widow, never having been to Sweden and not knowing any of her husband's relatives, remained in Spain, educating her children as good Spaniards. One of them eventually became an Admiral of the Spanish Navy. Today, a descendant of his, Louis A. Bolin, is the President of the Spanish Tourist Association. Two other sons of John Bolin went to Finland and Russia. The older one, Charles, settled in St Petersburg (now Leningrad) where he met the daughter of the foremost jeweler at the time.

He married her, and in this way entered the jewelry business. The firm he built up always made only gold in St Petersburg, leaving silver to their later Moscow workshop. His father-in-law only had two daughters; his two sons-in-law continued the business. Bolin's brother-in-law, an Englishman, was not interested in this profession and in his place Bolin's brother, Henrik C., grandfather of the present owner of the firm W. A. Bolin in Stockholm, became a partner in the company. The latter, in addition to his other artistic abilities, was also a singer. This brought him in contact with the leading composers of his time, like Rubinstein and Patius, and they dedicated several works to him. Through this music he became friendly with Tsar Alexander III; he was invited to give command performances at the Imperial Opera and was awarded a medal for his singing.

Henrik C. Bolin established a branch office for his jewelry in Moscow, then a very rich commercial city. He soon had a staff there of thirty silversmiths, with many more goldsmiths and sales people. As the Russian Imperial Family was related to most of the reigning Houses of Europe, the Bolin jewelry company came in close contact with the Courts of Great Britain, Germany, the Scandinavian and Balkan countries. Many personal friendships followed, resulting in an animated exchange of letters. These letters were kept in Bolin's private safe, but during the Russian revolution, when the firm was plundered by the communists, they disappeared along with the whole stock.

What especially gave the firm the prestige they enjoyed were the large orders received for coronations and royal weddings. These royal gifts were invariably designed by Bolin himself. The Imperial Crown Jewels, which were kept in St Petersburg, were also entrusted to his care. For each coronation and major royal function they were brought to Moscow by special train. Bolin always travelled in this train, remaining with the jewels in the Kremlin until the ceremony was over, and then returned with them to St Petersburg.

Each coronation and royal wedding was the culmination of years of work for Bolin. For instance, for the jubilee of Romanov in 1912, three years were needed to execute the gifts of the Tsar to other royal houses and to members of the Russian nobility: the ateliers and workshops of the firm were occupied all this time by these tremendous pieces alone.

In addition, the Russian nobility themselves often placed extremely lavish orders. The book by Grand Duke Alexander, *Once a Grand Duke*, describes the wedding jewels received at his marriage to Xenia, sister of Tsar Nicholas II; the silver for this stupendous wedding was also made by Bolin!

> *Jewelry: a pearl necklace consisting of five rows of pearls, a diamond necklace, a ruby necklace, an emerald necklace, and a sapphire necklace; emerald and ruby diadems, diamond and emerald bracelets, diamond breast ornaments, brooches, etc. All jewelry was made by Bolin, the best craftsman of St Petersburg. It represented, no doubt, a stupendous outlay of money, but in those days we judged the jewelry by the beauty of its design and colours, and not by its value.*

The Bolin family had a status in Russia that was unique. None of its members ever accepted Russian citizenship – they remained Swedish subjects. In spite of this, they held positions of high confidence in the Russian Empire. Henrik C. Bolin, as later his son William A. Bolin, was President of the committee that worked out the laws governing the Russian Mint. Several times the attention of the Tsar was called to the fact that a foreigner held this important position. Both Alexander III and Nicholas tried to influence the Bolins to become Russian citizens, promising them knighthoods. The Bolins, however, would not yield. One group of cousins to the jewelers finally accepted and received a title, but the family of Henrik C. Bolin remained Swedish. Their official excuse was that they preferred to be good Swedes rather than bad Russians, that they could never be considered true Russians as long as their name did not sound Russian and they were not members of the Greek Orthodox Church.

Before the First World War, W. A. Bolin, as the firm was now called, had a branch office at Bad Homburg an der Höhe, in Germany. The Imperial Family used to spend part of the summer there and the Tsar wanted his Court Jeweler to be with him so that he would be in a position to give presents to his friends. When the war started, the firm had to remove its assets from Germany: William A. Bolin collected his stock and brought it to Sweden. On arriving in Stockholm he showed part of it to the Minister of Foreign Affairs at the time, Knut Wallenberg, who proposed that he should leave the goods in Sweden, where Wallenberg would furnish elegant premises in one of the buildings he was erecting next to his bank, the Enskilda Banken. Bolin was not prepared to decide immediately, however. The thought of having Sweden not only as a place for his holidays pleased him, but for a while the plans were laid aside.

Wallenberg told the King of Sweden, the late Gustav V, that Bolin was considering the possibility of establishing a branch office in Stockholm. Gustav V knew Bolin from his visit to Moscow for the marriage of his son to the Grand Duchess Maria. He called for Bolin, invited him to establish a workshop in Stockholm, and promised to pronounce him Jeweler to the Court of Sweden. Wallenberg and Bolin discussed possibilities, and Bolin decided to open a branch office on the premises provided by Wallenberg. In 1916 the king inaugurated the new showrooms. Subsequently, the firm moved to larger premises on Sturegatan in the centre of Stockholm, where they still have their main showroom.

William A. Bolin brought with him from Russia many craftsmen, goldsmiths and silver-smiths, engravers, and chasers. Gustav Gedda was one of the best, a splendid engraver, who stayed with the firm till he was 80 and whose grandson is the famous Nicolai Gedda. They began to train Swedes to reach their high standards. Until import restrictions and exchange difficulties raised too high a barrier, Bolin supplied the fashionable jewelers on Rue de la Paix and Place Vendôme in Paris with the highest quality portion of their stock, such as cigarette and powder cases in gold and silver, often inlaid with enamel. Such trade is dwindling. The workshop on the edge of Stockholm now employs less than a dozen craftsmen.

Today, the Swedish Crown Jewels are examined by the present head of Bolin, Henrik C. Bolin, each time they are displayed. Henrik used to design all the jewelry and much of the silver, having studied in various ateliers such as Couvin Frères of Paris. The firm makes grand presentation pieces: among these was the gift from the Swedish people to Princess Ingrid at her marriage to Crown Prince, now King, Frederick of Denmark, an immense plateau with bowls, candelabras, platters and vases in sterling silver, one of the largest orders executed by any firm in modern times.

Hans Bolin himself, born in Austria and married to a Dutch girl, is now manager under his uncle Henrik Bolin and his aunt Mrs Kingston. Their able jewelry and silver designer is Barbro Littmark. She was born in 1918, qualified from the Konstfack in 1941, and has stayed with Bolin ever since. With them, she has made the firm famous for enamelling, often using pretty peasant designs from the province of Dalecarlia. The firm probably handles goods of a more precious nature than anyone else in Scandinavia. But there, as everywhere, and particularly after twenty years of socialist government, the trend is towards less luxury and more turnover. Bolin are becoming increasingly international. In 1966 they exhibited silver and jewelry by young British goldsmiths sponsored by the Worshipful Company of Goldsmiths; the King of Sweden and a Warden of the Worshipful Company contributed to nation-wide publicity against which the actual sales seemed small. Such is the present state of the highest quality trade. Selling the best work is always a difficult struggle and Bolin are committed to it.

32 3 pitchers
DM Bolin, Russia *c*. 1910

32

33 Large bowl (detail opposite and front cover)
D Joddli, a French sculptor
MO Bolin, Stockholm 1916

34 Silver and blue enamel prize for the best aviator in the air forces of Sweden, Denmark, Norway and Finland, whose badges are shown.
D Oscar Brandtberg
M Bolin 1931

35 Casket given by the Swedish people to King Gustav V on his birthday 1858–1928
DM Bolin, Stockholm

36 Traditional Russian late 19th-cent. cutlery made by Bolin in Stockholm *c.* 1920 with dies brought from Russia. The blades made at Eskilstuna.

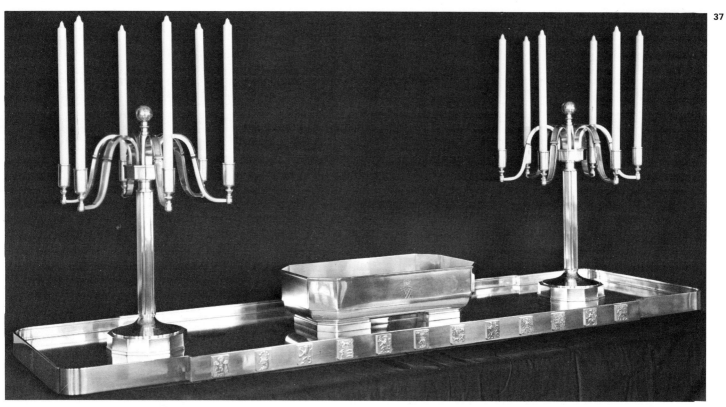

40

41

42

37 Suite given by the Swedish people
to the King of Denmark on his
marriage to the Swedish princess
Ingrid (daughter of the present
King of Sweden). The tray is
2 metres long and was jeopardized
by a high wind when carried to
Parliament over the bridges of
Stockholm.
D Oscar Brandtberg
M Bolin 1935

38 Award gold medal New York
World Fair 1937
D Oscar Brandtberg
M Bolin

39 **D** Brandtberg
M Bolin 1940

40 Box, lid enamelled with peasant
designs associated with the
province Dalecarlia.
D Henrik Bolin
M Bolin 1964

41 Bowl decorated with waste strips
from cutlery stamping. Given by
Bolin to the Worshipful Company
of Goldsmiths after their exhibition
'Young British Goldsmiths' in the
Bolin shop 1966.
D Barbro Littmark
M Bolin 1966
O The Worshipful Company of
Goldsmiths

42 Communion set with green and
violet enamel.
D Barbro Littmark
M Bolin 1961

GAB, Sweden

Most commercial firms which prove soundly based are also secretly based: like an iceberg they present an attractive pinnacle for their public face, but they keep in private the main bulk of their activity. The Scandinavian reputation for modern design was well earned, and Scandinavian people certainly have good taste. But nobody would claim now, any more than they would have done fifty years ago, that all the products of GAB, the biggest silver factory of northern Europe, are always distinguished, or even always new in design. Most people think of silver as decoration, and this means ornamental, decorative, frilly, and all the other words which a modern designer uses in a slighting way. But civilized life has always required ornaments, and artists who think the plainest work is best for silver sometimes end in bankruptcy.

Guldsmeds Aktiebolaget has, since 1907, employed two life-long staff designers of outstanding integrity, Jacob Ängman and Sven Arne Gillgren, and they have each been given freedom by the company to work on their own, as well as using the factory studios. Both have brought with them enormous prestige, and one could hardly imagine Swedish silver without the association between them and the firm. But a large proportion of GAB production is in traditional style using dies and tools which may be a hundred years old, more or less accurate replicas of antiquity. The same is true of the four largest factories in Europe today, Christofle in France, WMF in Germany, and the British Silverware group in England: the public everywhere wants a relief from the austerity of economical modern architecture. Old-fashioned, fussy designs can still be mass-produced today because the tools already exist. Austere modern design, the natural result of machinery, has only a limited appeal, though it too is easy to produce. It is elaborate modern design like, for instance, some of Gerald Benney's silver, which cannot be mass-produced because it is too expensive.

GAB started in Stockholm in 1867; by 1883 there were 150 workers. In 1900 a large building went up on the site of the present factory and in 1917 it merged with Guldvaru AB in Malmö, a big jewelry company started in 1845 with a staff of about 200. In 1961 they both joined Hallbergs, the biggest chain and retail stores in Sweden for jewels and smallwork, which also had workshops. The last merger took place in 1964 with AB Gense of Eskilstuna, bringing the total staff of the group to about 1500, much the biggest production unit in northern Europe.

Gense was founded in 1856 by Gustav Eriksson at Eskilstuna, to produce stove shutters and ventilators. Fifty years later he was already making nickel-plated and silver-plated tea and coffee sets, and silver-plated cutlery. But the firm became prominent only in the 1930s, when awareness of stainless steel spread from England, and led to a demand for 'something new' in tableware. In 1961 Gense launched 'Chef' – silver-plated stainless, but it is the solid stainless cutlery of Folke Arström which makes Gense immortal: for twenty years, since the war, his stainless steel patterns, 'Thebe' (1944), 'Facette' (1951), and 'Focus' (1955) have startled the world with their grace and elegance. This was the first cutlery to express the combined and unadorned needs of the new materials nylon and stainless steel, as harnessed by the creative designer, the producing machine, and the eater's manipulating hand. Lighter than any cutlery before, more precisely engineered and smaller, these pieces introduced a new delicacy to the table. Arström was Gense's head designer from 1940–64; his monument is the astonishing grace he gave to stainless steel. Copying and Japanese competition forced Gense's merger with GAB. But the companies remain distinct. Pierre Forssell has joined Arström and with his 'Pirouette' designs may prove to be the creator of the future as well as the publicist of the present.

Jacob Ängman's name is something of a legend in Sweden. Tall and serious minded, he influenced generations of students at home and abroad. He lived through and largely influenced the evolution of style from flamboyance to reserve, leaving his native silversmithing in a state of restraint from which it has not yet released itself.

He was born in 1876. The details of his life by modern standards are hardly spectacular. In 1893 he was a student at the Stockholm Technical School; from 1896–8 he worked at Otto Meyer bronze casters, again studying in the Technical School part time till 1903; in 1899 he spent four months in the Stockholm bronze foundry, C. G. Hallberg; in 1903–4 he studied in Germany working at Otto Bommer, metal sculptor in Berlin; he worked with Th. Müller, jeweler in Weimar under the direction of Van de Velde, and Brandstätter in Munich; from 1904–7 he worked at Elmquist of Stockholm, managing the engraving casting and metalwork there, and in 1907 he joined GAB where he remained till his death. In 1909 there was another three-month tour to Copenhagen, Hamburg, Paris, Munich, Nuremberg, Dresden and Berlin; in 1914 he took part in the Baltic exhibition; in 1919–20 he worked at St Peter's Church, Malmö with Gunnar Asplund; 1921, a further three weeks' tour to England – London, Birmingham, Oxford, and to Norway – Bergen and Oslo; in 1925 two weeks in Copenhagen and Paris, with the Paris exhibition; 1928, exhibition in the Metropolitan Museum, New York, and six weeks' travel to Copenhagen, Utrecht, Amsterdam, Cologne, Munich, Vienna, Prague, Leipzig, and Berlin; 1931, exhibition in London; 1935–40, worked with Folke Arström at GAB where, in 1937, Sven Arne Gillgren joined them; 'Rosenholm', finished in 1933 and released in 1935, in silver, is still Sweden's best-selling flatware, thirty years later, and is mass-produced by GAB; in 1942 he died and the National Museum, Stockholm gave him a memorial exhibition.

He was later than Georg Jensen and less individual. Perhaps it is for both these reasons that aesthetes have rated him higher, and the public lower, than Jensen. Jensen was the pioneer, Ängman the perfectionist; but, in retrospect, Jensen products are by far the more important, and it is Jensen's firm, not GAB which has devoted itself entirely to new designs.

Ängman's admirer and successor is Sven Arne Gillgren who was born in Stockholm in 1913, graduating from the State School of Art in 1936. He studied engraving in the C. G. Hallberg workshop in 1933, and like Ängman he has identified himself with GAB, from 1942 being the head designer in all its several undertakings. In 1958 he was elected a member of the distinguished Swedish Society of Industrial Designers. The same year he won first prize in the Stockholm Handicraft Association silver competition; in 1960 he won the important international competition for a new flatware pattern organized for the International Silver Company by the Museum of Contemporary Crafts in New York. In 1961 he won the award 'Swedish form, good form' and in 1963 the Prize of Honour from the Swedish Design Centre. In 1964 King Gustaf Adolf VI awarded him 'The Royal medal for distinguished artistic work, the Medal of Prince Eugen'. In 1966 he won the Eligius prize, the Prize of Honour of the Swedish Jewelers' Association, for artistic work in the goldsmiths' field. He has travelled far more widely than Ängmann. Though there are now many more designers of this calibre in the world and he is consequently less famous than Ängman, he has exhibited more often: about 80 exhibitions in Sweden, Britain, the USA, Canada, Germany, Switzerland, Norway, Italy, Holland, Austria, France, Belgium, Yugoslavia. He has made plate for more than one hundred Swedish churches as well as for Finland, Africa, Canada, and the United States. He is represented in many museums: in the National Museum and the Scandinavian Museum, both in Stockholm, the Röhss Museum of Industrial Arts, Gothenburg, the Gävle Museum, the Värmland Museum, the Ostersund Museum, the Sundsvall Museum, the Oslo Museum of Industrial Arts, the Nordenfjeld Museum of Industrial Arts, Trondheim, the Museum of Fine Arts, Houston, Texas, and the Museum of Contemporary Crafts, New York, USA.

It may be, however, that Gillgren will be remembered even more as teacher than as designer. There are some 1200 students in the Stockholm Konstfackskolan, about 15 a year passing out of his department. There is a technical trade school for Swedish silversmiths near Lynköping, but it is to Stockholm that the aspiring designers go, and it is to Gillgren's sympathies that they turn. Sweden will probably never regain the silver initiative which she once almost took from Jensen, and which an Englishman may now feel belongs in London. It is individual designers who now cause the pendulum of taste to swing, no longer national movements. But Gillgren's influence will always be towards a fine sobriety and that will always be a credit to his country.

43 Vasa pattern
   D Jacob Ängman
   M GAB, Stockholm 1921

44 Rosenholm pattern, after 30 years still the best selling cutlery made by this big firm.
   D Ängman
   M GAB 1935

45 **D** Ängman
   M GAB 1907

46 **D** Ängman
   M GAB 1918

47 **D** Ängman
   M GAB 1919

48 **D** Ängman
   M GAB 1924

49 **D** Ängman
   M GAB 1934
   This series of fine tea and coffee
   pots shows the evolution of
   Ängman's designs from the
   curves of *art nouveau* to the
   angles inspired by cubism

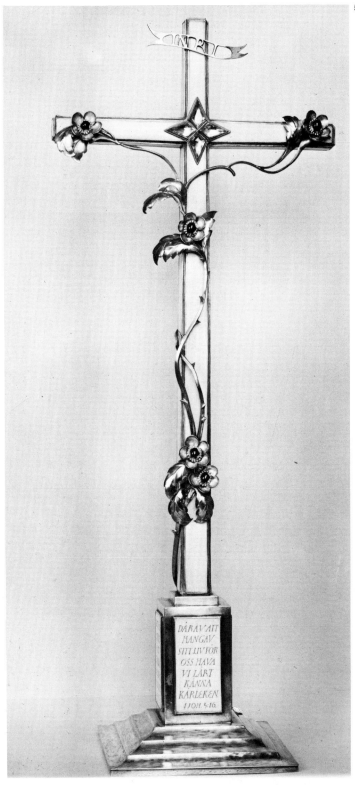

50 Altar cross, Margareta Church, Oslo: silver, ivory, rubies and aquamarine
D Ängman and Professor L. I. Wahlman 1925
M GAB

51/52 Chalice, St Petri Church, Malmö: silver, details engraved
D Ängman 1930
M GAB

53/54 Communion and altar sets for medieval church at Åtvid: silver, gold, rubies, aquamarine, amethysts, chased decoration
D Sven Arne Gillgren 1946–7
M GAB

55 Altar set for Nynastamms Church: silver and rock crystal
D Gillgren 1954
M GAB

56 Altar cross for Köping Church: silver and rock crystal
D Gillgren 1956   **M** GAB

53

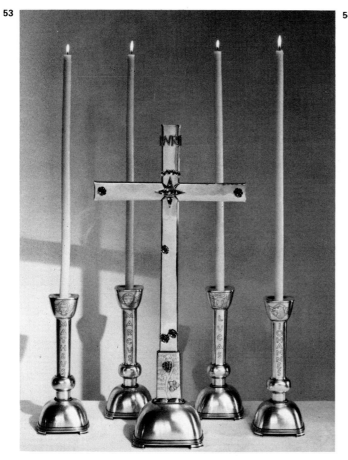

54

55

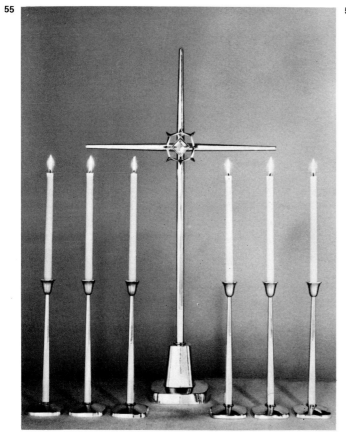

56

57

57 Altar set for St Andrew's Church, Slottstaden, Malmö: silver and rock crystal
   D Gillgren 1958–9
   M GAB

58 Flatware for Swedish embassies: silver
   D Gillgren 1946
   M GAB

59 Thebe pattern: stainless steel.
   D Folke Arström 1944
   M Gense, Eskilstuna

58

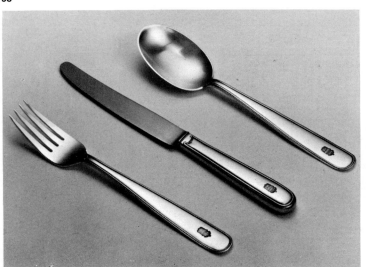

59

60

60 Altar set for medieval church at
Wä: silver on black granite socles
with rock crystal centre of cross
D Gillgren 1965
M GAB

61 Focus pattern: stainless steel
D Arström 1955
M Gense

62 The winning sterling silver flatware
in the world competition sponsored
by International Silver Company
of USA
D Gillgren 1960
M GAB

61

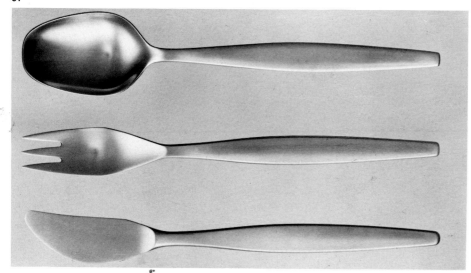

62

POTT, Germany

The world's four great steel towns were Pittsburgh, USA with no cutlery, Sheffield, England with light cutlery as well as heavy foundries, Toledo, Spain with its myriad tiny scissor mills, and Solingen in the Ruhr, much the smallest of the four, where knives were the main concern. Since 1945 vast new factories have been set up for instance in Ireland or in Japan, and the ancient pattern is changing. Newly invented machine tempering and polishing of blades is usually not so good as the old hand finish, but its cheapness is making it more and more common. The smaller workshops are disappearing and the bigger factories are mostly producing worse work in order to make more money. A shining exception is Pott of Solingen.

Founded in 1904, the firm made elaborate and unexceptional flatware until the advent of the founder's son Carl Pott. Born in 1906 in Solingen, he studied in the art school there for twenty-five terms, then in Schwäbisch Gmünd, then visited many of his firm's branches, resolving to banish from them the ponderous Edwardian baroque work that he saw everywhere. He was rewarded as early as 1937 with the first of the firm's astonishing series of international awards: a diploma of honour in the Paris World Fair, a silver metal at the 7th Milan Triennale 1940. Thereafter his firm – still family property – remained steadfast in its aim of graceful modern quality; its means were sterling silver (92·5 per cent pure), silver (80 per cent pure), 'Alpaka' plate with 90 gr. of silver plated on the surface, and stainless steel with 18 per cent chrome and 8 per cent nickel; and its achievement was international acclaim. Pott did for cutlery what Braun did for plastics or Olivetti for typewriters. He showed that the best can succeed.

The world has acclaimed Pott cutlery more consistently than any other modern flatware, and the firm has not been coy or mean – they have exhibited everywhere and competed with everyone. The list is impressive: it indicates, incidentally, how many chances there are in our self-conscious times since the war for a producer to make himself famous in the prestige section of the myriad trade fairs. In 1940 Pott won a silver medal at the 7th Milan Triennale, not yet the international yardstick it soon became. In 1951 there was a diploma of honour at the 9th Milan Triennale; in 1952 a commendation in the Geneva food and cooking design show; in 1953 a gold medal in the Düsseldorf modern design exhibition; in 1954 a gold medal at the 10th Triennale; in 1957 a silver medal at the 11th Triennale and a bronze medal at the Como display 'Colour and Form at Home'; in 1958 at the Brussels World Fair a gold star; in 1959 at the Sacramento California State Fair a gold medal; in 1960 at the 12th Triennale a gold medal; in 1962 at the Austrian Wels Fair a gold medal; in 1963 at the Sacramento California State Fair a gold medal; in 1965 the State prize of the government of North Rhine/Westphalia; in 1965 a commendation at the 1st Ljubljana Biennale for industrial design. Pott was in the Chicago 'Good Design' selection and, of course, is in the Museum of Modern Art in New York. If this cutlery perhaps lacks the absolute grace of Folke Arström's for Gense, or the grand weight and colour of Gerald Benney's for the Ionian Bank, it is good and sensible and generally useful. Some people handle spoons like pencils, others like garden spades, but everyone likes Pott.

Carl Pott was interested in the Werkbund – a German Design and Industries Association – founded in 1907 and revived in the fateful year of Hitler's accession in 1933. The Werkbund was convinced about modern design. This is one reason why Carl Pott abandoned the old styles he had inherited; another is his enterprise in using several designers besides himself. Josef Hoffmann, for instance, born in Pirnitz in the Vienna woods in 1870, still designing *avant-garde* cutlery in 1958, co-founder of the Wiener Werkstätte and the Deutscher Werkbund; or Elisabeth Treskow of Cologne whose research into gold granulation in the 1930s was so useful; or her student, Alexander Schaffner of Basle, for whom Pott produced a cutlery pattern in 1960. Then there is Professor Wilhelm Wagenfeld, of Bauhaus fame, who was designer at Weimar between 1919 and 1924 and who won a diploma at the 1951 Triennale, or Dr Gretsch the well-known industrial designer with whom he worked. The Pott team are always claiming ever more simplicity and ever greater fitness for purpose. But the truth is more complex: the human hand can do anything, the mouth does not mind the shape of utensils. It is the eye which matters, and to the eye this cutlery is both stimulating and harmonious.

63

64

63 Stainless steel 1956. Gold medal
Brussels World Fair 1958.
D Dr Josef Hoffmann, Vienna
M C. Hugo Pott, Solingen

64 Diploma of Honour, IX Milan
Triennale; Brussels World Fair 1958
Gold Medal. Silver 'Alpaka' plated
D Dr Hermann Gretsch
M Pott 1951

65 Gold Medal, X Milan Triennale
Silver 'Alpaka' plated
D Elisabeth Treskow
M Pott 1954

66 Gold Medal, X Milan Triennale;
Gold Medal, Brussels World Fair
1958. Silver 'Alpaka' plated
D Carl Pott
M Pott 1954

65

66

67

68

67 Diploma of Honour, IX Milan
Triennale. Stainless steel
D Carl Pott
M Pott 1951

68 Silver Medal, XI Milan Triennale:
the only distinction for German
cutlery there
Gold medal Brussels World Fair
1958. Stainless steel
D Carl Pott
M Pott 1957

69 Gold Medal, X Milan Triennale
Stainless steel
D Carl Pott
M Pott 1954

70 Gold Medal, X Milan Triennale.
Chosen for Good Design Show,
Chicago, and Museum of Modern
Art, New York. Stainless steel
D Carl Pott
M Pott 1954

69

70

71

72

71 Gold Medal, Passau 1962
   Stainless steel
   D Carl Pott
   M Pott

72 Gold Medal, XII Milan Triennale
   Silver 'Alpaka' plated in 1967
   Lufthansa had 3 million Pott place
   settings in use
   D Carl Pott
   M Pott 1960

73 Gold Medal, XII Milan Triennale
   Brussels World Fair 1958
   Silver
   D Josef Hoffmann
   M Pott

74 1965 State prize of the
   government of North Rhine –
   Westphalia. Shown at Montreal
   World Fair 1967. Stainless steel
   D Carl Pott
   M Pott

73

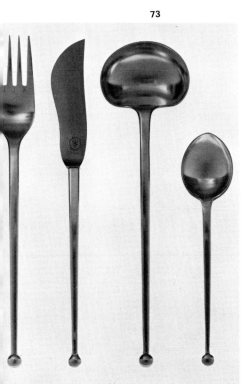

74

75 Bowls in silver 'Alpaka' plate, or
   stainless steel. Gold Medal XII
   Milan Triennale 1960.
   DM Pott

76 Shown at Montreal World Fair 1967
   Silver or 'Alpaka' plated
   D Carl Pott
   M Pott

Württembergische Metallwarenfabrik, or WMF, Germany

This vast metal goods factory at Geislingen now employing 6500 factory and office workers, was founded in 1853 by Daniel Straub (1815–89), a miller who had been repairing tools and railway construction equipment. Employing sixteen men, he opened his 'plaque factory', called Straub and Schweizer, producing tableware from silver-plated copper sheets. The increasing well-being and rising standard of living of the German middle classes after the 1870 Franco-Prussian war meant that Straub had chosen the right moment to introduce his plated ware – solid silver was too expensive and other manufacturers had failed to find an adequate substitute. The healthy demand enabled him to cut his prices, introduce electroplating, and head towards mass-production, his initial aim. By 1866 the firm employed 120 workers.

Straub associated his factory with A. Ritter & Co. of Esslingen (founded in 1871 and already making silver-plated articles), making one company in 1880, employing 500 people. Straub passed the management of the enterprise to Carl Haegele, who had been technical director of Esslingen. Haegele (1848–1926) successfully expanded WMF, making connections with firms abroad and increasing exports. The expansion meant new buildings, and a very advanced staff welfare scheme. In 1883 Haegele added a glass works to the existing plant. He modernized all the machinery, and in 1897 WMF took over Schauffler & Safft of Göppingen, a light holloware factory. Hans Schauffler (1852–1904), the head of the new addition, succeeded Haegele, who resigned, and then in 1904 a board of directors took over, led till 1939 by Hugo Debach (1871–1939).

In 1923 a minor boom followed the invention of reinforced 'points of wear' by additional silver deposit, and soon after, the introduction of 'Cromargen' stainless steel 18/8 alloy tableware brought further strength and variety to the company. Research into the nature of patina on metals led to the multicoloured 'Ikora' metal, a new fashion for coloured crystal called 'Myra' and 'Ikora' glass. The department of galvano plastic electrotype reproduction and of bronze casting laid some heavy weights on the world: the statue of liberty at Lesbos, a solid silver altar in Malta, a cupola on the Royal Palace at Bangkok, and the Ghiberti Baptistery doors in Florence, copied and new in the firm's museum. Under some managements such a varied enterprise would have rambled to its ruin, but WMF maintained its impetus as Europe's largest silversmith. Branch factories were built in Berlin, Cologne, Vienna and Warsaw and exports even as far as East Asia exceeded home sales. But the two world wars devastated the business.

Today there are ninety WMF shops in Germany, there are 5500 different products including coffee machines and glass, exports to eighty countries, and there are branch factories at Hayingen and Laichingen in Württemberg, at Hausham in Bavaria, and at Volos in Greece. The 1966 sales were 248 million Deutsch Marks. Such size and diversity should encourage the creative artist-craftsman: there will always be a public reaction against the anonymity of giants. But for the many small factories employing a hundred men rather than thousands, WMF is a threat. The answer lies where WMF themselves found their success, in specialized invention and thorough technique; or where Jensens found theirs, in an unmistakable artistic character.

77 Pewter inkstand
   D Beyschlag
      c. 1900
   M WMF
   O Stadtmuseum,
      Munich

78 **M** WMF *c.* 1900
79 **M** WMF *c.* 1900

80 Vase
   D Beyschlag *c.* 1900
   M WMF
   O Stadtmuseum, Munich

81 Pewter inkstand
  D Beyschlag *c.* 1900
  M WMF
  O Stadtmuseum, Munich

82 **D** Wilhelm Wagenfeld 1951
  M WMF
  O LGA, Stuttgart

83 Vegetable dish: silver
  D K. Dittert 1967
  M WMF

84 **D** Dittert 1967
  M WMF

82

83

84

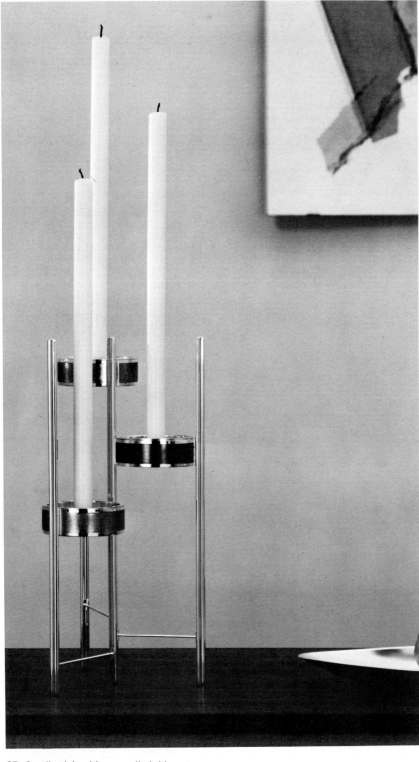

85 Candlestick with enamelled drip
   pans in several colours: silver
   Ht 50 cm
   D Dittert
   M WMF

86 Three candlesticks with coloured
   glass: silver
   D Dittert
   M WMF

86

FABERGÉ, Russia

Just as designer-craftsmen are sometimes considered to be in the trade, and sometimes to be artists outside it, so Fabergé himself falls between two stools. His firm sometimes made silver of their own inimitable and rather useless sort; and they sometimes stooped to wearable jewels. But the little golden empire they created for themselves was one that could only exist in highly sophisticated court societies in times which we now call decadent. Small precious objects of ornament or 'virtu' were much sought after in the Italian Renaissance and eighteenth-century Paris and Dresden; and they are popular even today in New York where Jean Schlumberger of Tiffany's designs delicious trivia for the table. These precious nothings are so fragile that one hardly dares touch them, so small that they almost escape notice, so easy to steal that they often live in storage, and so unadventurous artistically, that they give almost universal pleasure. They are made to give pure joy in exquisite work, and people often succumb to them when they cannot face reality.

Fabergé's family was French; his father, Gustave Fabergé (1814–93), founded a jeweler's business in St Petersburg in 1842, and prospered, no doubt because he was a foreigner, appealing to the Russian nostalgia for the cosmopolitan life of western Europe. In 1870, his son, the famous Peter Carl Fabergé (1846–1920), took control, and in 1882 was joined by his younger brother Agathon. In 1884 the firm received the Royal Warrant of the Tsar Alexander III, and embarked on the famous series of Imperial Easter eggs. The family achieved great success through their original standards of craftsmanship: in 1882 a gold medal at the Pan Russian exhibition, Moscow; in 1885 a gold medal at Nuremberg; in 1888 a special award at Copenhagen. In 1900 Carl Fabergé won the Légion d'Honneur after the Paris World Fair, and there were many other awards. Branches were established in Moscow (1887–1918), Odessa (1890–1918), Kiev (1905–10), London (1903–15).

Modern aesthetes may mock at Fabergé's small talent; *art nouveau* passed him by, and he is the perfect example of craftsman, not artist. But he had the impertinence to realize it, doing no more than refine and re-interpret eighteenth-century Dresden for which he fell as a student. In his tiny way Fabergé is great, as the facts of commerce both at his time and today prove.

He had a wonderful range of coloured enamels, often working at heats of 700° and 800°C instead of at the usual 600°C, often building layer upon layer, a process of amazing patience and refinement. He unearthed many forgotten semi-precious stones, using particularly the heavy greed jade from Siberia called nephrite. A complex Fabergé treasure such as may now be seen in the fine collection at Luton Hoo in England, or at the Lillian Thomas Pratt Collection, Virginia Museum of Fine Arts in the United States, might be the result of the conference of all the specialists involved in making it, and might take one craftsman earning perhaps five roubles a day (ten shillings or $1.50) as long as fifteen months to finish. The old Russian custom of giving presents and eggs within the family at Easter, intoxicated the tsars. The first egg of 1884 was so popular that the Imperial Family ordered new eggs each year. Vice-Admiral Sir Francis Pridham, in his book *Close of a Dynasty*, recalls that he was given one by the Empress Marie when she fled from Yalta aboard his ship in 1917; he noticed that the ladies of the family wore necklaces of miniature Easter eggs for six weeks before the Ascension. One such necklace of enticing delicacy is in the collection of Lady Zia Wernher at Luton Hoo.

In 1910 Fabergé lost his test case against the Worshipful Company of Goldsmiths. He disputed the British hall-mark law, claiming that his products should be exempt as works of art, that sterling silver – the minimum British legal standard – could not satisfactorily be enamelled, and, anyway, that his work, being made in Russia, must not be damaged in the London assay office where repair was impossible. The Worshipful Company used in its support expert evidence from the famous English *art nouveau* designer Alexander Fisher. Worse was to come. The Moscow shop was closed by the Bolsheviks in 1918, and Fabergé died in Lausanne.

At the height of their fame Fabergé had a staff of over 500 with some thirty designers, but Carl and Agathon supervised all work closely and themselves designed the great majority of pieces made in the St Petersburg workshops. Perfectionism is a feature of aristocracy. Fabergé was not interested in the new popular taste for repetition and uniformity. For him detail was all, far more important than the lust for money of our modern age.

87 Serpent clock egg, given to Marie
Feodorovna by Czar Alexander III,
1885–90. 4 colour gold, the egg
enamelled in royal blue on a
guilloche ground, Louis XV style.
Signed by the workmaster Michael
Evlampievich Perchin who, till his
death in 1903 when he was
succeeded by Wigström, made the
whole series of magnificent
Imperial Easter eggs.
DM Fabergé. Ht 7¼ in.
O Wartski

88 Strut clock, silver enamelled; on
Fabergé's standards, a typical
humble domestic piece.
Workmaster Michael Perchin,
c. 1900
DM Fabergé. Width 5¼ in.
O Wartski.

87   88

89

89 Scent bottles
top left: striated agate with
granulated gold mounts with
carved laurel swags and a
cabochon ruby finial signed by H.
Wigström and bearing gold mark
72. In original case. £350
top centre: Russian gold mounted
rock-crystal bottle with ruby snap,
the top decorated with a cat's eye
and diamond cluster. £180
top right: Gallé glass *art nouveau*
bottle of a mulberry colour with a
samarodok gold mount relieved

with carved leaves and cabochon
ruby berries and diamonds. Signed
by Friedrich Köchli of St
Petersburg. In original case. £210
front left: scent bottle enamelled
translucent pale blue over a
wave-patterned field with two-
colour gold carved laurel mounts
and a rose diamond set in the top.
Signed K. £125
front centre: varicoloured agate
scent bottle with hammered gold
mount set with a cabochon ruby
and diamond spray. Signed by the

early workmaster Erik Kollin. £375
front right: scent bottle in carved
topaz quartz formed as a pine cone,
the carved gold mount set with
diamonds and a large carved ruby
of fine colour by Bolin and bearing
the initials of his workmaster W.
Finnikov. This piece, fitted into a
Fabergé holly-wood case, appears
to have been part of Fabergé's
stock, a procedure not unknown
in those days. **DM** Fabergé *c.* 1900.
O All Wartski Ltd, whose 1967
prices are given.

TIFFANY, New York

Tiffany's achieved a hundred years ago what is almost impossible today; they became the
leading firm in New York. Ever since then they have had a name which carries them over
weak patches, but which they have, from time to time, strengthened with success. They are
now manufacturing and retail silversmiths and jewelers.

Charles Lewis Tiffany (1812–1902) founded the firm in New York in 1834 to sell household
goods, including slices of the first transatlantic cable at a dollar each; he started manufacturing jewelry in 1848. The first name was Tiffany and Young, a partnership, changed in
1853 to Tiffany & Co. Almost all silver was bought from John C. Moore who had begun manufacture in 1827, and whose son Edward soon joined him. Tiffany soon absorbed all the Moores'
production. In 1868 Tiffany's and Moore's were incorporated: Edward C. Moore became the
designer as well as being a director. He was a collector of oriental *objets d'art* and knew the
dealer Samuel Bing of Paris. His work shows Indian and Japanese influences, giving a glamorous sense of distant experience to the firm, which it has always since maintained. His initial
was stamped on all silverware (M), along with 'Tiffany & Co': since then the president's initial
has been stamped on all the firm's silver. In the 1850s and 1860s the firm first introduced to the
USA the British standard of sterling silver: 92·5 per cent pure. Tiffany's were very successful
in international exhibitions and were greatly admired abroad, particularly in France.

90 (*l. to r.*) Antique Chinese carved
mother-fat jade converted by
Fabergé into a powder compact
with red gold bezel and translucent
green leaves. Workmaster's initials:
Henrik Wigström. Diam. 2¾ in.
Aventurine and gold écuelle,
handles with garnets and emeralds
on white enamel ground. Initials:
Henrik Wigström. Diam. 5½ in.
Rhodonite box with translucent
green emerald sides, diamond
thumb-piece. Bonbonnière, drum
from Imperial Porcelain Factory
with cypher of Nicholas II and
Romanoff double-headed eagle.
Initials: Henrik Wigström (1911)
ht 1⅛ in.
  DM Fabergé
  O Wartski

91 Miniature table in Louis XV taste
by Carl Fabergé, coloured gold,

lapis lazuli top, underside enamelled
pale yellow over a guilloché
background. The blue stone and
yellow enamel together represent
the Rothschild racing colours;
Fabergé designed several other
objects in this colour scheme for
Leopold de Rothschild. Signed and
initials of the chief workmaster
Michael Perchin, ht 3½ in.

92 Box in nephrite, the jade
particularly loved by Fabergé.
  DM Fabergé
  O Wartski

93 Jewel casket, bronze and enamel
  D Victor Prouvé *c.* 1893–4
  O Musée de l'école de Nancy,
      Nancy, France

94 Box gold and lapis lazuli
  D Jean Schlumberger for Tiffany
      & Co. New York 1965

The factory had been at 49–55 Prince Street, New York. When it was removed in 1897, the floor boards were burned and refined and yielded no less than 1300 ounces of fine silver and 600 pennyweights of gold. The new factory was built in 1897 at Forest Hills, Newark, New Jersey, where it still is, and the machinery installed was said to be the first for a silver factory to be run by electricity anywhere in the world.

Louis Comfort Tiffany was born in 1848. Designer and maker of glass, bronze, silver and jewels, a product of America's millionaire class, he changed the taste of a nation, intoxicating his country with *art nouveau*. As son of Charles Tiffany, the smartest jeweler in town, he had a good start. Tiffany's were the first American silversmiths to win an award from an international jury – at Paris in 1867, using the English 'sterling' standard – and by 1870 the Tiffany shop in Union Square claimed the world's biggest gem collection, with pieces from the Esterhazy, Marie Antoinette and the French royal family. The customers, Havemeyers and Goulds, Astors and Vanderbilts, wanted art with individuality, show even if it was vulgar, above all ornamental richness, and Louis Comfort had the talent as well as the human understanding to provide it.

He studied painting with George Innes and Samuel Colman, in 1869 worked in Paris, then visited and loved the exotic Near East and Tangiers; in 1870 he was in Cairo. After his second trip abroad in 1875 he became fascinated by decorative art, and the current revival in England under Ruskin and Morris, Liberty and Whistler. His father's shop, opened at 221 Regent Street in 1868, must have given him access to the new oriental fashions in London. Later with S. Bing's 'L'art nouveau' in Paris it was to provide his main European sales outlet. His real career began after the 1876 Philadelphia Centennial Exhibition when with Innes, Colman, and La Farge he started the Society of American Artists (he was Treasurer in 1878), and the Society of Decorative Art, from which, in 1878, he formed Louis C. Tiffany Company, and Associated Artists. He was encouraged by the silversmith and collector Edward C. Moore, his father's partner in Tiffany and Company. He already knew the important people and big commissions for interior decoration came soon and easily: the Madison Square Theatre 1879, the Library of the 7th Regiment New York Armory 1880, part of the Union League Club 1881, the Lyceum Theatre, the state dining room, the red and blue rooms at the Washington White House 1883, as well as many private homes. The *Studio* in 1897 referred to Tiffany's eclecticism as a 'decorative jungle'.

In 1875 he started on glass at Thills Glass House, Brooklyn, and his first decorative window in what came to be called 'American' glass was put into St Mark's Episcopal Church, Long Island in 1878: these richly-coloured suggestive patterns quickly became popular. With Andrea Boldini, a glass-blower from the Salviati factory at Murano, he started his own glass-making firm at the Heidt Glass House, Brooklyn. As early as 1880 he patented the first of his beautiful inventions; the iridescent 'favrile' technique, the result of exposing hot glass to metallic fumes and oxides. This type of glass was first suggested by the excavations of glass from Pompeii and Rome, Troy and Mycenae. In 1873 Ludwig Lobmeyer showed it in Vienna, and in 1878 Thomas Webb were making it and showing it at London's Grosvenor Gallery. In 1893 Arthur Nash of Webbs went to run Tiffany's new glass-works at Corona, Long Island, thus taking with him the skill of one of England's leading glass pioneers; his two sons Douglas and Leslie, also joined the firm. Their 1893 Chicago exhibition chapel in a Byzantine style, all glass and no metal, was much admired, and in 1895 Bing asked Tiffany to make ten stained-glass windows for his new Paris shop, 'La Maison de l'art nouveau', designed by Bonnard, Vuillard, Toulouse-Lautrec, Serusier and others, followed by Frank Brangwyn whose work was shown in the Crafts Gallery in 1899. Tiffany had achieved a synthesis of fine and applied art which was sadly rare in the whole *art nouveau* movement: with it came a return on the huge financial investment which his experiments had required and which his family wealth had made possible.

The Tiffany Glass and Decorating Company from 1892 and the Tiffany Studios from 1900 till 1936 employed hundreds of female 'craftsmen' and made bronze and enamel cigarette boxes, vanity cases and desk sets, all hand-finished in a wonderful variety of designs. His lamps of all sorts were the latest fashion – petrol, gas and electric, made of bronze with glass shades. Thomas Edison patented the incandescent lamp in 1878; Mackmurdo had exhibited an electric lamp in the Liverpool 1886 exhibition, but it was Tiffany who worked with Edison and made the lamp a work of art – he won a grand prix at the Turin 1902 exhibition for a lily cluster typical of this organic form. These products were mostly either signed or numbered but the system used is so obscure that Tiffany glass can seldom be dated, and is sometimes confused with its competitors and later imitators.

He acquired almost limitless wealth, became Art Director of Tiffany and Company, and concentrated on designing heavy jewels, 'personalized jewelry', after his father's death in 1902. He made an Englishman, Joseph Briggs, director of the glass-works, whence Briggs acquired the fine collection now at his birthplace in Accrington, Lancashire. Tiffany built the amazing Laurelton Hall at Oyster Bay, Long Island (1902–4, burned down 1957), costing over $2,000,000, in which he entertained his artist's colony. He gave extraordinary parties; at one in 1914, peacock was the main course and as an appetizer girls in classical drapes carried live peacocks round the table on their shoulders. In 1911 he produced the famous mystical curtain of glass panels still in use in the Mexico City opera house.

This all takes some reconciling with his statement that 'simplicity . . . is the foundation of all really effective decoration', and his declared admiration for William Morris. In fact, unlike Morris, he was creator rather than missionary, and cared for his products more than for any theory of fitness for purpose or idealistic socialism. But in 1918 he set up the Louis Comfort Tiffany Foundation to which he gave Laurelton Hall, and he spent his later life enjoying the company of the young artists who got fellowships to study there, and whose work he sometimes made up. It was with this philanthropic effort that he perhaps came nearest to realizing his claim to be the Morris of the twentieth century. By 1932 the Tiffany Studios were bankrupt and in 1933 Louis himself died. Family wealth had made the whole giant enterprise possible, and perhaps it was this same wealth which sapped the necessary creative vitality away from Tiffany's declining years.

Since 1918 New York had become a shop window for the world: not only Jensen, but Cartier, Van Cleef and Arpels, Buccellati, and Harry Winston, set up their smart Fifth Avenue shops, and all the big department stores assessed New York as the best barometer for trade. Competition was killing, but the potential gains limitless. Tiffany's responded to the challenge as most old firms do; they became exclusive, declined to advertise, and retreated from the market place.

In 1956, however, Tiffany's grasped the nettle, and established a new life for themselves. Jean Schlumberger, the brilliant French jeweler, who still spent half his time in his Paris work-shop at 4 rue de Ponthieu, was lured to close his own New York business and become Vice-President of Tiffany's with a private room on the second floor in which his stupendous jewels are shown. Aware of the dazzling creative inheritance set up by Louis Comfort, the firm was now ready once more to produce new masterpieces, again to satisfy the impressive American urge for fine art patronage. Schlumberger brought Tiffany's back to a high level of art. His exhibition at Wildenstein, New York, 1961, and his pieces shown in London at the Goldsmiths' Hall international exhibition the same year, proved his extraordinary genius. From 1963 he has increasingly concerned himself with gold boxes and table ornaments, designing, for instance, a series of miniature blackamoors such as one might find in a sixteenth-century Venetian *palazzo*. Many great heiresses are now proud to own a Schlumberger piece. His success is partly due to good craftsmanship not always found in New York – small workshops who make for him, such as Louis Feron, are few and far between – partly to his own horror of repetition. He does not like other people copying his designs, but still less does he like to copy them himself.

95

95 Scent bottle: glass, gold and
   diamonds
   D Louis Comfort Tiffany
   M Tiffany & Co, New York *c.* 1900
   O Anthony Harwood

96 Teapot, silver enamelled etched and
   gilt, part of a 4 piece service given
   to the museum in 1897 by 'a friend
   of the museum'. Length 11 in.
   DM Tiffany & Co.
   O Metropolitan Museum of Art,
   New York

96

Training and technique

Rules and models destroy genius and art.
William Hazlitt (1778–1830), *On Taste.*

Until the nineteenth century, art was rightly considered as something one was born with: if one had such an extraordinary talent one probably became a misfit, impossible socially, bankrupt financially and incapable of everything except dreaming. One's training concentrated on life drawing and technique generally, with the aim of getting inspiration from, if not actually copying, the old masters. Sculpture and painting were all important and the great national academies and institutes held almost unquestioned sway in matters of fashion and even of sales. Puccini's *La Bohème*, Somerset Maugham's *Of Human Bondage*, or Flaubert's *L'Éducation Sentimentale* tell the story quite accurately, even though it now seems to us wonderfully picturesque. It differed very little from the days of Michelangelo who learnt sculpture 400 years ago under Lorenzo dei Medici in Florence, except in one important respect: there was now a barrier between fine art, which was inspiring, and applied art, which carried no message and was therefore negligible.

Almost all practical learning was centred round the workshop, factory or atelier, an intimate but restricted setting which usually worked very well; the student or apprentice would find his own master, and extract from his manufacturing workshop experience such information as he needed. The training which produced most of the great nineteenth-century craftsmen would seem to us quite inadequate. No national comparative standard of examination; no national professional bodies, such as institutes of architects, to ensure that the public were not defrauded, and that architect-designed bridges and buildings did not collapse; not even any widespread general education to allow potential artists to show their talent. Genius somehow broke through, though nobody knows how much better and more lively art life might then have been if art education had kept pace with the increase in population and industrial activity.

At the beginning of the nineteenth century there were only two museums in England – at London and Oxford, and they were both usually closed to the public. State patronage of art depended on the erratic benevolence of the sovereign – private patronage at public cost. By 1865 there were twenty-three public museums and art galleries in Britain. Art education was born at last.

Following the 1851 exhibition in London, the world's first great universal display of manufactured goods, Prince Albert and many others tried to raise the status of applied art in relation to painting and sculpture and founded what is now the Royal College of Art in London (as well as the Victoria and Albert Museum). This was one of the first and most successful of all the art schools where every art subject is taught. The Stockholm 'Konstfack' – Konstfackskolan, now probably Europe's biggest and best design school, with over one thousand pupils from almost every country, was started in 1844, the smaller Pforzheim school in 1877. The Oslo Museum of Applied Art was founded in 1876. Nearly all these enterprises were paid for by State or city; even the rare privately financed schools, like the Ruskin school of drawing beside the Ashmolean Museum in Oxford, or the Barcelona Escuela Massana in the splendid medieval Hospital of Santa Cruz, founded in 1929 by an art-loving pastry-cook, are now nearly always state subsidized. It would be tedious to list the dates and sizes of all the leading schools, but the same sort of consideration applied everywhere. There had been too much romance for painting and sculpture, too much contempt for applied art and trade.

Starting in England, the feeling gained strength that applied art mattered. In the Birmingham art school in 1932 there were already 4600 students, 310 of them specialized silversmiths. The word 'design' was hardly recognized and 'mass-produced' was still a distinctly dirty label in aesthetic circles. But money speaks, and the big factories were winning a social status for provincial cities and for the new merchants, which threatened and eventually conquered the ancient court life of the capital cities. Governments realized the need to harness machinery, to bridge the gap between art theory and industrial practice, and during the last half of the nineteenth century ponderous new art school buildings in the Gothic or Renaissance style were built all over Europe; London, Birmingham, Sheffield, Edinburgh, Pforzheim, Hanau, Schwäbisch Gmünd, Solingen, Copenhagen, Stockholm – these were the centres of modern silversmithing. The cities had been created by the skill of workshop-trained apprentices over the centuries. During the last one hundred years or so, they were helped by systematic courses in technique and design in their art schools, all of them monuments to their own period, none of them yet suspecting the rapidly approaching dominance of the machine. Beside them, the great exception was the Glasgow school, the overpowering masterpiece of Scotland's greatest architect Charles Rennie Mackintosh. Here was extraordinary inventive vision coupled with productive capacity. But the whole dream depended on one man's brilliance, and there was no basic idea to perpetuate after his departure.

It was in 1918 that Walter Gropius created the Bauhaus, the most famous milestone in the history of art education. The old Weimar Academy was combined with Van de Velde's old Weimar school of arts and crafts from 1919–25, in 1926 moving to Dessau. Walter Gropius, the principal, himself designed the building which attracted an increasing number of admiring international pilgrims until the Nazis killed it in 1933. Its archives survive at Darmstadt, in brilliant clinical contrast to the earlier ornate Jugendstil buildings by Josef Olbrich and others for the art colony there.

Until the Bauhaus, art education aimed to nourish the magic of individual imagination and to equip this with as much technical ability as the artist-craftsman needed to create unique pieces in his studio. At the Bauhaus team work became normal; fine and applied art were at last united. Each student had two teachers for each subject, one artist and one craftsman tutor, and machines abounded. Students were to design for mass-production, and if divine inspiration had to be tempered to fit a lathe, the result would have social validity. Often a designer would be only one of a group, which would probably make the product appear anonymous and therefore give it a new quality of non-controversial strength. Bauhaus designs were austere and functional, and would probably have seemed to Morris or to Ruskin, as they certainly did to the old *beaux-arts* academies, unspiritual and therefore undesirable. Naum Slutzky, the Bauhaus jeweler, felt nostalgically that the most delightful jewels had to be unique and therefore hand-made; but he spent much of his Bauhaus time trying to tune in to the organization's theme, evolving modules and mass-produced units which could be assembled in so many different ways as to suggest handwork and personal style. With such artists as Paul Klee, Marcel Breuer, Kandinsky and Moholy-Nagy, the Bauhaus was an amazingly accurate foretaste of modern times – very daring fine art, very geometrical applied art, and almost no hand craftsmanship.

Since 1945, silver training has settled into four channels. One can be a traditional apprentice, studying at the bench under contract perhaps for five years, learning the varied manual skills which used to be understood in the phrase 'fine craftsmanship'. The old methods are still the best for the old results, and today in London, for instance, at the Sir John Cass College, some six hundred apprentices from the highly skilled manufacturing workshops come for one day a week, having been released by their masters to broaden their manual training experience. A similar scheme, teaching technique almost separate from art, works at the trade school near Linköping in Sweden. As Chaucer said six hundred years ago, 'The lyf so short, the craft so long to lerne'.

Secondly, one can acquire a very high degree of skill supervising some complex, perhaps revolutionary, and probably semi-secret machine, such as the spot-welding device used by J. & J. Wiggin Ltd at Walsall in Staffordshire, to fix handles and spouts to the 'Alveston' pattern teapot by Robert Welch. It is necessary to use such a machine as efficiently as possible, so that the money invested in it yields the largest possible number of teapots, and with all metals, but particularly with stainless steel, polishing and cleaning are expensive and wasteful, so that welding must be accurate and clean. If the jet aircraft engine is the true modern expression of 'fine craftsmanship', the best craftsman is no longer the one who is most versatile with his hands, but the one who masters the workings of an intricate machine most thoroughly, so that he can use it to produce whatever variety of effects his designer wants. Training for such an operation may be specialized and may take only six months or a year, compared with the five years needed to master the full range of ancient hand skill with hammer and punch.

Thirdly, one can study one's own craft for three or four years in an art school. In 1966 there were 164 British art schools with 25,000 students. This, nowadays, is usually an uneasy compromise between drawings on paper and metal on the anvil. Until 1939, most art schools, if they taught crafts, concentrated on technique: students learned how to make a teapot from beginning to end with their own hands and it was hoped that God had already given them the intuitive imagination to create it in the mind's eye. Hidden in this admirably optimistic scheme, there were several flaws: hardly any students learned to make teapots at an economic speed. Consequently when they set up their one-man artist-craftsman studios they went bankrupt. They found they could only make a living by teaching at another art school, from which other students would get into a similar predicament. Furthermore, art schools used to accept pupils who had not got the necessary visual imagination and whose motive for learning was simply that they enjoyed using their hands. Some of the most stimulating and successful practising artists had the impertinence to train themselves: 'self-taught' often appeared in artists' biographies. The best talent was often outside the schools. In fact the art school almost seemed to be a self-perpetuating parasite on the edge of modern creative art, instead of its mainspring.

Perhaps the most damaging aspect of this scheme was the isolation of the factory: art school training took little notice of industrial production, and it was usually possible to complete a three- or four-year course in silversmithing without ever going inside a big cutlery factory. Pupils mapped out their futures in wholesome artist-craftsman studios, not in squalid industry. Conversely, and quite naturally, factory managers tended to find art school people unpractical and over-opinionated. The art schools were set up to make a marriage between art and industry but they nearly completed a divorce.

The fourth training possibility is the study of the new science of industrial design. Since 1945 there has been a radical change; the machine is no longer ogre, but god; the craftsman and hand-crafts are considered sentimental and silly; classical perfection is the aim rather than individual expression, and the wild swings of fashion which have always been a feature of civilized life are wrongly considered irrelevant to the needs of modern society. One of the first and most distinguished historians to suspect that England could in fact design, Professor Nikolaus Pevsner, called his pioneering book of 1937 *Industrial Art in England*; but thirty years later there is really no such thing as industrial art. Design is the word. Even the Society of Industrial Artists wants to change its name. Logical design, not whimsical art, is what manages machines, and it is in machines that our hopes for salvation lie. Industrial design is to be the means of improving our environment.

Art schools have mostly decided that the idea of technique is out of date – noting, for instance, what effect varnish may have on coloured canvas, or how long a spout is necessary to contain tea in a teapot; nor is invention any longer so important. Instead, a student must have a well-organized mind, able to coordinate all aspects of a design problem – materials, price, manufacturing, market, and appearance. The industrial designer has to organize the craftsman who has himself usually given his place to a machine. In many schools these ideas lead to too much study and too little nourishment. For the small craft industry, courses in 'basic design' or 'basic form' have found an uncertain target: the factories are too small to be able to employ highly-qualified paper designers. But as the world becomes more monotonous, there is more, not less, demand for small scale original production. The intimate knowledge of technique, and some specialized manual skill in silver and jewelry should, therefore, still be a vital part of our school training.

Ways of making silver vary enormously. Hand-raising is the slowest and often still the best, and in most artist-craftsmen's studios one can see its attraction. A flat sheet of silver is hammered into a depression on a large wooden block, hence the name of this stage of the operation, 'blocking'; the sheet is then turned over and hammered from what will be the outside of the bowl on to a specially shaped steel anvil or stake. The process from flat sheet to completed cylindrical coffee pot, for instance, will need thousands of hammer blows, and may take two or three weeks. Gerald Benney's achievement in raising a large jug from a flat sheet in four days for the Worshipful Company of Goldsmiths' film, *A Place for Gold*, was compared with the speed of lightning, and the strain of it gave him jaundice for six weeks after.

A very common alternative method is spinning: a shaped wooden chuck, or sometimes a succession of chucks, is rotated in a lathe. A sheet of silver, also rotating, is bent over the surface of the wood with a long steel lever. It is very difficult to spin thick metal, so spinning is normally associated with very light goods; furthermore, the shape of the bowl cannot be adjusted on a lathe with any sensitivity, so spinning often means a hard and obvious outline. Finally, the metal naturally tends to get thinner as it is stretched round the edges, and spinning has no effective remedy against this weakness.

The third method of shaping silver is by stamping – the sharp blow of one hardened steel die descending onto another, usually having been dropped from a height by a very heavily weighted device like a pile-driver, worked by a rotating belt in the ceiling. Each object will need several blows; there is the constant risk of crooked placing and the noise is literally deafening. The same effect is achieved in big hydraulic presses, relatively silent and enormously strong, but much more expensive to install. The main problem is the initial cost of steel dies – in 1967 for a new cutlery pattern this may be as much as £10,000 or £15,000. They can partly be cut by drills or routing machines, often operated from a plaster model with a sort of pantographic remote control mechanism, but the finishing of a die has to be done by hand and it is a difficult craft. Production will in future, no doubt, increasingly be by the new method of pressing called 'flow forming', using a single 'female' die, and in place of the male an oil fluid, which forces the sheet of silver into the cavity beneath, sometimes with the added encouragement of suction from a vacuum. Apart from economy in dies, this method of shaping, which is sometimes refined to use a sudden explosion and is hence called 'explosion forming', surmounts the difficulty of overhanging surfaces into which only expensive collapsible dies of the conventional type can penetrate.

Casting is the fourth and probably the oldest means of shaping, and is particularly suited to coarse work. A wooden or metal pattern is pressed into the surface of a small box of fine wet sand of a rare consistency found in only one or two parts of the world, such as Tripoli in North Africa. The upper half of the pattern is impressed upon another box of sand; the pattern is removed, the two faces of sand are united and molten metal poured down tiny channels or spines. The silver produced from each sand mould will probably contain some air bubbles and some traces of the unevenness in the sand, and it will certainly be thick and heavy because molten metal will only run easily through a wide cavity.

The new casting method commonly used for silver – centrifugal casting – only works on a relatively small and light scale; the mould is all plaster. It is fixed to a wheel which rotates at high speed by electric motor; the molten metal released at the centre of the wheel flies outward under pressure and, therefore, fills the cavity with great accuracy.

Assembly usually involves soldering: a rod of silver or gold of a lower melting point than the main object is melted under a gas flame onto the seam to be joined. This is usually done on a pile of charcoal, itself often on a small revolving table so that the heat penetrating underneath the piece is held and thrown back onto it and distributed evenly. The solder itself then runs freely round the whole crevice to be joined. In a small workshop the equipment used for soldering is often the same as for annealing, the regular softening under heat made necessary by the metal's compression while it is worked. In a factory, pieces are put on a metal belt which carries them through a sort of gas or electric annealing oven.

Then there may be engraving, a pattern cut out of the surface of the metal with a sharp spike, or chasing, a pattern pressed into the surface with hammer and punch. All these processes can be used with varying degrees of suitability on gold or silver or the various base metals normally used for plating: gilding metal, electroplated nickel silver (EPNS) or German silver as it used to be called, brass or copper and even stainless steel, though it is a brave man who chooses solder rather than welding for this. If the object is now to be plated, it will first be cleaned in an acid vat, then hung in a solution. The electric current dissolves metal into the solution from the big slabs through which it passes, and deposits metal from this solution all over the object to be plated.

These are the main production stages, often decentralized so that the head factory in an old centre like Birmingham may do no more than assemble parts which it buys from specialized out-works, one for spouts, one for handles, one for plating. The last hurdle, the polishing, is unexpectedly the highest. Many schemes for mechanization have been tried, from the rotating barrel filled with sand, spoons and forks, to the regimented knife blades on carborundum strips. But there is, as yet, no substitute for the human eye, so the best finish is always still obtained by hand. As polishing inevitably causes dirt, with oily sand and rouge and revolving mops, and as it is an uncreative job it has, like washing up with which it is almost comparable, proved as unpopular as it is necessary.

Sometimes polishing, first with rough brushes then with soft mops, finishes the job – and nearly every manufacturer will say he could produce more if he could get more polishers. Sometimes burnishing or flat hammering is necessary too: to make a flat dish really flat, after the embellishments have distorted it, comes a last round with a big hammer at a flat stake. This is often followed by burnishing by hand with a polished agate which is rubbed over the surface to seal its pores, or it may be stoned to remove the hammer marks.

Such is the range of technical and human material with which the designer must grapple. Small wonder if art schools are indecisive. They cannot fix the right accent for their training because the industry itself cannot decide whether it is mechanics or craft, whether it is at the beginning of a new epoch or the end of an old, whether it is alive or dead.

97 Exercises in basic shape by
Alexander Petrini, student at
Stockholm's Konstfackskolan 1967.

97

98

99

100

98 Bowl with experimental texture by
Birger Haglund, from Crete,
student at Konstfackskolan.

99 A contrast in texture.
DM Olle Ohlsson, a brilliant self
taught originator 1965.
O Nordiska Company,
Stockholm.

100 Spice boxes, design fantasies by
Bengt Liljedahl, student at
Konstfackskolan.

101

102

Silver chalice, enamelled foot, ht 20 cm. Students naturally try to develop their visual imagination by creating personal designs. The only way of testing a bright new idea is to make it. One can then decide whether the beauty of the shape, as with this foot, is more important than any possible inconvenience of balance or, as here, any difficulty in cleaning.

DM The class of Karl Dittert, Werkkunstschule at Schwäbisch Gmünd. This is the national centre of German silversmithing where perhaps 5000 men earn their living in the manufacturing industry, mostly in small workshops. There may be as few as six factories there like Bauer, employing about 200 men each. The other German metalwork centres are at Hanau for miscellaneous wares, including precious jewels, and Pforzheim, where there are some 30,000 people working in the lighter jewel workshops and factories, and 90,000 inhabitants. In all three towns there are fine art schools, as well as at Düsseldorf and elsewhere.

102 An ingenious way of giving interest to a tube: repeated pierced and applied sections emphasized with black oxidisation. The shapes are harsh, and the black will wear off.
DM Neil Harding while a student at the Royal College of Art, London 1965.

103 Condiment sets in glass and stainless steel, entries for the competition held at the school for Eisenberg-Lozano Inc. of New York, 1965. Such competitions give students a rare opportunity suddenly to become famous, but unless the objective is clear the results are often bizarre or undistinguished and therefore useless, and the sponsor's high costs may not be repaid simply in terms of free editorial publicity given.
   DM Konstfackskolan, Stockholm 1965, where imaginative art training is given. The big Swedish technical school for apprentices is not in Stockholm but near Linköping.

104 Melting at Tiffany's (Murray Hill, New Jersey factory)

105 Rolling silver sheet at Tiffany's

106 Stamping cutlery blanks at Tiffany's

107 Spinning at the Taunton, Massachussetts factory of Reed & Barton, probably the oldest surviving US silversmiths. The sheet is coaxed over a chuck usually made of lignum vitae, the hardest wood. Compared with handraising the form is less subtle: it cannot be adjusted as it progresses. The metal is thinner – thick metal is too stiff to spin. And the density of the metal is less well suited to its final use – only hammering can harden the metal just where the greatest stress will occur.

108 The first process for hand-made silver: from inside on a wooden block a bowl is 'blocked'. Then from outside on steel stakes it is patiently 'handraised' till the flat sheet reaches the required shape. Gerald Benney in London making an alms dish for Coventry Cathedral 1965.

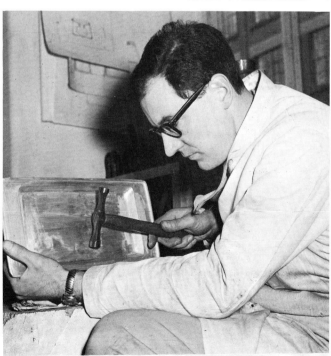

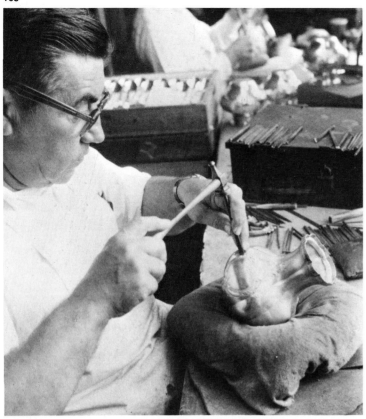

109 Chasing – lines are impressed into the surface of the metal with hammer and punch. Reed & Barton, USA.

110 Engraving – cutting into the surface of the metal with a sharp tool or graver. Gerald Benney making a coffee pot for the Ionian Bank, London

111 Soldering – metal with low melting point flows between the surfaces to be joined, often lubricated with borax flux. Annealing in a small workshop may look similar – the object which has hardened as a result of repeated hammering or working, is softened under a flame, the degree of softness achieved depending both on the amount of work still to be done and the rigidity required of the finished piece. Sven Arne Gillgren with a student in the Stockholm Konstfackskolan

112 Polishing – a notorious bottleneck in silversmithing throughout the world, because the work is inevitably dirty with oily sand and rouges, and it is less well paid than the highly skilled hammering jobs. No machine has yet been invented as an adequate substitute for hand polishing. The balance between polishing thoroughly whilst not rubbing off too much of the precious surface, is most easily reached by hand. Reed & Barton

113 Polishing flatware can be entirely mechanized but not for expensive semi-hand-made pieces to whose irregularities only the hand can properly respond. Reed & Barton

114 Soldering – the pieces to be joined are clamped carefully in position and the bar of solder held ready to flow when the gas melts it. C. J. Vander, London

115 Cross
    DM Leslie Durbin. Given to Guildford Cathedral by the Worshipful Company of Goldsmiths, 1937, the winning design in an open national competition

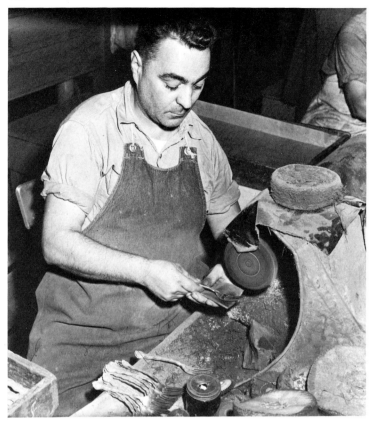

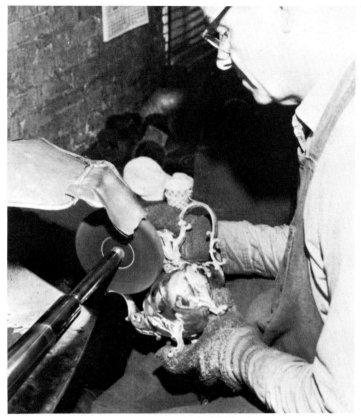

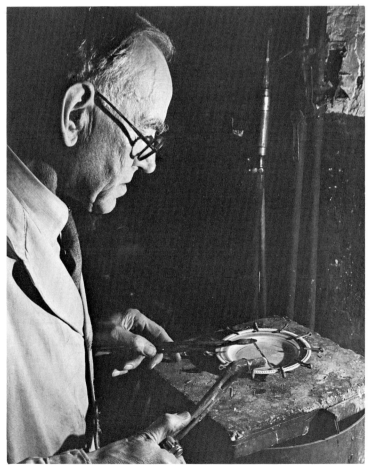

116 Rudimentary shine can be achieved by electropolishing – the object is vibrated at great frequency causing friction with the liquid in which it is suspended. Reed & Barton

117 The normal method of plating one metal onto another is by electro-deposition. (See Elkington, who invented the process in 1840, Christofle and Wiskemann.) By the old mercurial gilding an amalgam of gold and mercury is painted onto the silver surface which is then heated on a charcoal flame, the mercury volatilizing and the gold and silver surfaces fusing and mingling. Sheffield plate, invented by Thomas Boulsover in 1740 and killed by electroplating, is a sandwich of copper and silver.

Genesis and Exodus

The metals gold and silver, production, structure of the trade, display, exhibitions, publicity

The mechanics of industry are not mechanical in the baser sense, but full of fine philosophy.

W. A. S. Benson, designer in *Elements of Handicraft and Design,* 1893

God made nothing tidy.

Richard Jefferies, naturalist, in *Round About a Great Estate,* 1881

The apparatus of eating has eased our gastronomy but strained our language. Unglamorous but vital, cutlery ought to mean only knives. In the trade one says flatware, not cutlery, to describe knife, fork and spoon place setting together, as opposed to holloware for pots and all things hollow. But laymen like whatever sounds right, and for most people cutlery is a manageable idea whereas flatware is obscure technical jargon. Cutlery is the word most often used to describe all the implements of eating.

In the Middle Ages silver was normally covered with a thin layer of gold; we now call it gilt, but it was then called gold, and the people who made it goldsmiths. Thus the London medieval guild which has always included silversmiths and jewelers as well as workers in solid gold, was and is called the Worshipful Company of Goldsmiths. Tindale and other early translators of the Bible, four hundred years ago, used the words goldsmith and silversmith indiscriminately. Silversmith is a new word dating from that time, and silver gilt is newer still.

Plate used to mean gold or silver or anything precious made for use on the table, probably because rich men usually ate off silver plates or silver plates plated with gold. The word comes from *plata*, the Spanish for silver. It was Sheffield Plate, invented in 1740 by Thomas Boulsover, that caused the etymological inexactitude, because it was relatively non-precious, intended as a cheap imitation of solid silver; the situation worsened in 1840 with Elkington's invention of electroplate, meant to be a very cheap imitation of the already cheap Sheffield Plate. The modern word plate now has no universal accurate meaning, but is normally just an abbreviation for electroplate, the normal trade form of which is 'electroplated nickel silver'.

Distinguished scholars, however, like Sir Charles Jackson or Mr Charles Oman, still use the word in its old sense: their great books on antiques: *A History of English Plate* (1911), and *English Church Plate* (1957) are about only solid metals. So are E. Alfred Jones's *Old Oxford Plate* (1906) and *Old Plate of Cambridge Colleges* (1909), *Old English Plate* by Cripps (1901) and *Greek and Roman Gold and Silver Plate* by G. E. Strong (1966).

So the word plate is now almost useless for logical definition, simply serving to stimulate one's poetic sense. The whole subject of precious metals is indeed impregnated with poetry. Gold was a staple diet for Shakespeare, and has enriched literature of all times and places. The Mexican Aztec word for gold means 'excrement of the gods': The Peruvian Incas called gold 'sweat of the sun', and silver 'tears of the moon', and identified gold with the sun. Everybody knows what Lorenzo told Jessica in *The Merchant of Venice*: '. . . look how the floor of heaven is thick inlaid with patines of bright gold', and everybody has yearned at some time with Richard III for 'Wedges of gold, great anchors, heaps of pearl/Inestimable stones, unvalued jewels!'

Such is our romantic nostalgia today that we always imagine gold and silver treasures to have been made only in the past, and we think of new gold simply as money. As Lord Keynes said, gold is now 'dug up in South Africa, only to be buried again in Fort Knox'. The 'mad magic of gold' is undeniable if indefinable. Avarice may now be a bigger incentive than art, but both have certainly contributed to the fantastic gold rushes on the discovery of new deposits. The excitement in Dawson City, the centre of the Klondike gold rush seventy years ago, was such that half the adventurers reaching this remote spot never bothered to look for the gold even though they had suffered incredible hardships to get there, walking and climbing through the snow. They sank back, their imaginations sated before they ever touched the metal.

Gold came mostly from Brazil and South America in the eighteenth century, and in the nineteenth from Russia. The world's outlook was transformed by the frenzied 1848 rush to California, launched by the discovery at Sutter's Fort, shortly before Mexico made the worst ever bargain, selling upper California and New Mexico to the United States for only $15 million. During 1849 over 100,000 prospectors are said to have arrived from all over the world, perhaps half of them travelling 3000 miles across the continent by horse, wagon or foot; the rest by sea. As many as 500 ships, deserted by their sailors in San Francisco Bay, were counted in July 1850. Small wonder! A single washer sieving by a river bed might get £50-worth of gold per day. The towns were glutted. There was no home life; only 8 per cent of the local population were women, in the actual mining areas only 2 per cent. But all this chaos was, in fact, a transformation scene. The output of California alone in the first five years probably exceeded that of half a century in the whole of Brazil.

In 1851 it was the turn of Australia. The population of New South Wales quickly doubled, and neighbouring Victoria was worried enough to offer a reward of £200 to the first person discovering gold near Melbourne. The challenge was met with the Ballarat field. In 1893, remote Kalgoorlie, where discoveries were made in 1888, became the centre of another rush for wealth. Output there reached its peak about 1900, but has now dwindled to a safe £7½ million yearly.

In 1898 Alaska responded to systematic prospecting; Dawson City, founded in 1896, sprang up at the junction of the Yukon and Klondike Rivers, and for ten years probably led the world's production. Its population quickly rose to 30,000 but already by 1901 it was calculated at only 9100, and by 1941 it had sunk to 1000. Most of the metal was recovered by washing, but the idea of industrial equipment to dredge for gold deposits deep down in the river bed mud and to crush rock brought up from mines was already born. There is today in Bonanza Creek, outside Dawson City, forgotten and deserted, a huge gold dredge as big as a cathedral, supposed to have sucked up some $8 million-worth of gold from the Klondike. From 1898 to 1905 as much as $100 million was mined from creeks in the region, but by 1907 it was over. The beds were scoured clean. Some relics are in the museum at Whitehorse.

Jack London wrote some of his poems in a cabin here which is still preserved and the whole amazing episode has been well recorded by him, by Robert Service and by Rex Beach. The word bonanza came from the colonial Spanish settlements by Los Angeles, and meant 'fair weather and prosperity'. It was first used in English by the Californian gold rush speculators to indicate incredible riches easily won. By the time of Dawson City the word was truly identified with mining and the Bonanza Bar there still stands as its memorial.

South Africa gave the world its fourth quite unexpected golden fortune in fifty years, the Witwatersrand. The Transvaal and Orange Free State area has produced by far the most gold the world has ever known; prospectors first discovered promising quartz there in 1884; in 1886, George Harris, an Australian gold digger and house builder, with his friend George Walker, detected the fabulous Main Reef, the foundation of the whole Witwatersrand industry.

Already by the time of the South African war, Winston Churchill was spellbound by the size of Johannesburg, the fruit of all this mining wealth on the arid plain, named after himself by the first Boer Minister of Mines, Christian Johannes Joubert; nearby was Klerksdorp, described by Arthur Conan Doyle as an 'insignificant little litter of houses'. The huge gamble continues, though it is now not small prospectors but industrial empires, such as the Anglo-American Corporation of South Africa, The Union Corporation, The Chamber of Mines and even governments themselves that place the stakes. Gold may first have been found at Klerksdorp by a hunter who shot a buck; the animal, kicking as it died, broke a piece of outcrop rock, revealing the glint inside. For years there were only shanties here, but in 1936 serious mining began and today the area around Klerksdorp claims to yield over 11 per cent of the free world's gold.

Gold is the mysterious buttress of the world's finances; in 1933, when the Americans again pegged their dollar to this metal, they declared 'it is a fetish.' But it is the one commodity which is limited in supply, easily portable, very small in bulk and universally desirable for its beauty as well as its value. Whatever economists may say, gold remains an indispensable background to paper currency, and the amount of public reliance on paper depends upon the gold behind it. What is always in dispute is not the dependence of currency on gold, even when it is not specifically tied to a 'gold standard'; it is the unanswerable question as to how much gold is necessary in proportion to the paper money in circulation to avoid a sudden crisis of confidence, such as that of 1929.

The British Chancellor of the Exchequer, James Callaghan, said in 1967 that solving the world's problems through gold alone is 'a caveman's solution' and John Ruskin noted ninety years ago that only 'partially savage nations' need gold for currency. In this sense, alas, we are all savages. In the nineteenth century about one third of the total gold output may have been used for industry and art; today the figure is much lower, perhaps one thirtieth or $50 million. In recent years annual western gold production (half of it from South Africa), together with Russian sales in the west, has averaged about $1500 million, that is, about the same as the whole world's production throughout the seventeenth century, and if stacked together as gold bricks, it makes about a 13-foot cube. Of this, official government gold bank reserves have absorbed only some $500 million annually. Nearly all the rest of the world's gold today, perhaps nearly $1000 million-worth each year, is privately hoarded – an estimated $210 million-worth for India alone, most of it smuggled.

During four and a half centuries, 1493–1940, the world's gold production, as recorded in mints, banks and counting houses, may have been about $42,000 million at today's price, or a 44-foot cube. What a strangely powerful metal that so small a bulk should determine so many destinies.

Silver shares much of the history of gold if not the glamour. It used to come from Bohemia, Germany ('Thaler' is the origin of the silver 'dollar') and Spain; and after the Spanish conquests in America, from Mexico, Peru and Bolivia. The most picturesque source was Mount Potosi in the Bolivian province of that name, whose metal was supposed to have particular softness and colour, so that some Spanish colonial silversmiths' pieces came to be known as 'Potosi Silver', one of the many picturesque and probably fictional associations which influenced early collectors. Even a Birmingham factory, now part of Barkers, once used the name. Many of the nineteenth-century developments of gold also offered silver as a by-product, notably the famous Comstock lode (1859) in the Sierra Nevada and the great Anaconda mine (1894), and the Australian Broken Hill deposits (1885). Today, silver comes mostly from Australia, the USA, and Mexico.

Most people equate gold and silver with the idea of treasure, and treasure with priceless antiquity; but prosaic coins, rather than ornaments, have probably always accounted for most of the world's production, and today the situation is still less romantic. These metals have unique electrical and physical properties, and are therefore consumed by the electrical and electronic industries in unprecedented quantity. The leading American refiners, Handy & Harman, estimated in their 1966–7 annual report that 150 million ounces of silver were consumed in the United States during that year for industrial purposes, an increase of nearly 10 per cent over the previous year. They guessed that the non-Communist world used 470 million ounces on industry and coinage, 'a very sharp drop from the all time high of 722 million ounces in 1965', a decrease due to the removal of most silver from the US coinage. Of this, industry used some 357 million ounces, and coinage outside the US some 54 million ounces.

Gold and silver are rather like sun and moon; complementary but distinct. Gold, being much rarer, is much more valuable and is therefore bought more often than silver purely as an investment. At the 1967 prices of the world's biggest bullion firm, Johnson Matthey of Hatton Garden, London, gold is about 252 shillings per ounce and silver about 11 shillings: gold prices may change rapidly with the economic situation, whereas silver will be more affected by industrial demand.

The gold and silver trade is, alas, in terms of quantity only an interesting survival on the fringe of this vast new potential. For a craftsman, the character of both metals is similar; both are soft and easily worked, both of them in the alloys normally used, harden as they are hammered or compressed, both therefore have to be regularly softened or annealed under heat at intervals during the working, and both have surfaces capable of being cut and banged into a wonderful variety of textures. Gold is very dense, too heavy to be practical for large objects that have to be moved, such as coffee pots or dishes. Indeed the most obvious distinction between the two metals, or between gold and silver gilt, is simply their weight. Lastly, pure gold is impervious to chemicals and therefore never tarnishes, never rusts or decays: hence the marvellous survivals from old tombs and underground deposits. Silver, on the other hand, despite all recent researches, does tarnish, and may for instance affect the flavour of wine drunk from it, or mark a dress if a bangle rubs on it. In all these respects gold is more extreme than silver, and they are both different from, for example, stainless steel, which is very much harder and of more uniform colour, or copper which can be much softer and which is always much more affected and discoloured by changes in atmosphere.

Perhaps the most bewildering technicality in silver does not apply at all to gold. It is 'fire stain', the blackening caused under heat during annealing or soldering, by the oxydization of the copper alloy near the surface. Some makers like to leave hammer marks, as the 'pedigree' of their production, on the surface, in which case their last operation would be to heat the piece in the open air, spreading fire stain all over it, leaving it with the rather darker patina of most antiques. Other designers prefer extreme precision and austerity and remove the stain with acid, or simply polish away the outer dark layer, revealing the white silver beneath. Many makers today, though usually those interested in quantity rather than quality, will leave the fire stain, but conceal it by electroplating the finished object, thereby putting a very thin layer of bright white silver on top of the thicker black stain. Both layers will of course eventually wear through. Perhaps the worst solution is to electroplate fire stain with rhodium, a hard bluish metal of the platinum group, giving an untarnishable finish almost comparable with chromium, a lamentable victory for utility over beauty. The problem arose only with the accuracy of modern refining: the usefulness of copper to harden the silver, to give depth to its colour and to stabilize it when polished, outweighs its nuisance value.

In the eighteenth century almost every maker would have his own small hand-rolling mill and often his own melting equipment too. Today he will buy from one of the big, highly scientific metal firms like Johnson Matthey or the Sheffield Smelting Company in England, Handy & Harman in the United States, Degussa in Frankfurt, Metalli Preziosi of Milan, Nyström Matthey of Sweden or Universal Matthey elsewhere, firms which will not only know the capacity of different alloys but also produce tube, bars and sheets in wonderful profusion.

So, if the maker is exceptionally big, like the new British Silverware group with its specialized Heeley Rolling Mills in Sheffield, the cutlery empire built up by the brilliant financier Mr Charles Clore, it may provide its own raw materials; but more frequently it will not. Many of the older factories still retain their crucibles and furnaces, relics of the days when they were proud to carry through the whole production operation from beginning to end, not realizing that this was both uneconomic and unscientific.

The factor common to nearly all British silver is the astonishing living tradition of the hall-mark, the small row of punch marks which for nearly seven centuries have guaranteed the quality of British gold and sterling silver. Without these marks it is illegal to sell gold and silver in Britain – they represent the oldest form of consumer protection in the world. Normally a piece of British sterling bears four punches. First, the maker's mark, usually his initials; second the town mark showing where the piece was assayed; third, the standard mark showing the quality of metal, probably the lion passant for sterling silver; and last, the date letter – when it was made.

The primary meaning of the word hall-mark is a mark applied at Goldsmiths' Hall (for centuries the headquarters of the London guild of goldsmiths) to denote the quality of gold and silver wares. Hall-marks refer only to the quality of the metal and do not indicate a standard of design or craftsmanship. They are only applied to wares which have been accurately tested or assayed. Goldsmiths are proud to know that the term 'hall-mark' has been extended, first to include the marks struck by other statutory assay offices, and later to mean any indication of high quality. But while the principle is easy, the details are complicated.

This 670 years old practice of hall-marking was first recognized officially in the reign of

King Edward I. A statute of 1300 provided that no ware of gold or silver should be sold until tested by the 'Gardiens of the Craft' and struck with the leopard's head. This was first known as the King's mark and was used alone until 1363. Later the leopard's head became the London town mark. Each assay office now uses its own alphabets, starting at different dates: as the leopard's head denotes London, Edinburgh has a castle, Sheffield a crown, and Birmingham an anchor. There were also assay offices at Chester, Exeter, Glasgow, Newcastle, Norwich, and York, but these six are now closed.

At Goldsmiths' Hall in London business is now enormous; about 100 people worked there in 1966 and in the three big months before Christmas they marked 2,100,000 gold objects and 525,000 pieces of silver. The total for the whole of 1966 was more than 4,100,000 objects, ten times the amount of 1954. More quantity here than quality perhaps, but nevertheless proof that public and precious like to go together.

The British system of hall-marks is unique. In all other countries there is either no special treatment for gold and silver – the public is protected from fraud in Japan or Italy, for instance, by general regulations comparable to the British merchandise marks act: the maker must not mark his products inaccurately. In the USA, the maker is required to stamp on to gold and silver the quality of the metal, and there are government inspectors who occasionally test samples taken from the shops. The loopholes here are so big that the system only succeeds if the manufacturer is honest. The commonest scheme is exemplified in Stockholm. The Swedish government assay office receives there all gold and silver, tests a small proportion – one or two of each series of objects – and, if these are up to standard, marks them all individually. The law in France and Germany is similar. Perhaps the most curious mark is in Denmark: it is optional. Some firms think it helps sales and therefore use it; others do not bother. Some, like Jensens, feel their own trademark is better known and find the hall-mark superfluous. Much of their product is in sterling (92·5 per cent pure silver), and the highest quality recognized by the Copenhagen assay office is the normal continental standard, 80 per cent pure; to use such a mark on sterling would be uncomplimentary. In each country and for each year the mark is different, a practical buyer's guide as well as a fascinating lure for collectors.

The structure of the trade during the first half of the century changed very little. In each big country there may have been two or three dozen artist-craftsmen's studios, like C. R. Ashbee in England, Louis Comfort Tiffany in New York or, rather later, Andreas Moritz in Nuremberg. Naturally they were extremely different from one another – it was their business to be so. But they all had in common a sort of crusading feeling that they were at least keeping alive, if not actively reviving, the noble spirit of fine craftsmanship. In the early years of the century, particularly in England, these studios often gave birth to small cooperative groups whose commercial success was usually, and sadly, negligible, whose artistic standing was often questionable, and whose life depended on charity, not on public interest or demand.

Ashbee's own commercial Guild of Handicraft, founded in 1888 and dissolved in 1908, delighted in the use of untrained labour whose incompetence was more than outweighed by their enthusiasm. Ashbee himself wrote, 'None of these men had any trade workshop experience; such experience was, in the eighties when we began our work, regarded rightly as rather a detriment.' Merchandise from the Guild was considered, at least by its own members, as superior to the semi-craft silver made by such firms as William Hutton in Sheffield and successfully sold all over the world as well as at Liberty's in Regent Street; trade was a dirty word.

There was the Sheffield Art Craft Guild founded in 1894 to which many of the designers and craftsmen there belonged, often using it as an outlet for their spare time creative energy, in contrast to their routine work in a factory. Then there was the Birmingham Guild of similar character and the Gloucestershire Guild which still lives on at Painswick, the last survival of Ashbee's art colony at Chipping Campden. In London, the Art Workers Guild (1884) and Arts and Crafts Exhibition Society (1887, now the Designer Craftsmen) were founded by William Morris, Walter Crane and others, to encourage artists and to sell their products.

In Finland there were two such societies formed during the 1870s: Suomen Taideteollisuus-yhdistys (Konstflitföreningen, Finnish Society of Crafts and Design) 1875, and Suomen Käsityön Ystävät (Friends of Finnish Handicraft) 1879. A craft school was founded also in the 1870s, but in 1875 it was reorganized as the Central School of Industrial Art, today called Taideteollinen oppilaitos (Institute of Industrial Arts). From 1875 it was under the management of Konstflitföreningen and was not an independent body. In 1966 it was given to the government. They all had the same aim: the 'promotion of the Finnish handicraft and its development in a national and artistic direction'. The ideas of William Morris and John Ruskin reached Finland in the 1890s. It was the painter Gallen-Kallela who brought the Swede Count Louis Sparre to Finland. He, in turn, founded the Iris factory at Porvoo-Borgå in 1897 and brought to it the English/Belgian artist A. W. Finch. Finch was an intimate friend of Van de Velde and actually persuaded him to abandon painting and go in for crafts and architecture. He was one of the founder members of 'Les XX' to which, amongst others, people such as Ensor belonged. The Konstflitföreningen keeps developing magnificently; during the last fifteen years it has organized over 150 exhibitions of modern Finnish design throughout the world. It is, in fact, the world's second oldest arts and crafts association.

The oldest, started in 1845, is the Slöjdföreningen in Sweden. This famous society, under its director Erik Wettergren, worked with Orrefors under Johan Ekman to show in the 1920s that cheap need not mean bad. Orrefors, themselves founded only in 1898, gave a healthy and almost indispensable push to Swedish crafts. The society flourishes as a centre of intelligent discussion and as a modern design pressure group, with a small annual Government grant of some £3000, and with beautiful show cases in the Central Park of Stockholm. Svenskform is the related permanent modern Swedish design centre, purely for exhibitions and not for sales, and there are fine shops like Hantverket and Konsthantverkerna which get no grant but manage both to make money and to promote young designers. Hantverket is very wealthy, owning the hotel Gillet with its enormous showroom beneath. The group started by selling and it has continued to do so: a small jury of three distinguished people, regularly changed, guides artistic policy, and weekly sales have always been considered the important criterion of success.

In 1918 the Norwegian applied art society, Foreningen Brukskunst, was founded. In Denmark too, there is the old Society of Arts and Crafts and Industrial Design, started in 1906, which has no permanent local showroom but which often supervises Danish promotions overseas, such as the great 'Arts of Denmark' exhibition which toured the USA from coast to coast in 1960. The Danish government gives about £12,000 for this society to participate in the regular Scandinavian Design Cavalcade, the summer festival which alternates from capital to capital. Den Permanente in Copenhagen, started in 1931, and partly managed by the society, is a most ambitious show-place for modern design, its aims being both to exhibit and to sell, with no distinction between handwork and mass-production; it is understandably criticized for being half-hearted in all these directions. It sometimes fails to make a profit – in 1967 its losses, up to £15,000 annually, were being paid by the Danish government. But it succeeds in the vital job which most shops never have the courage even to start. It is much the best, and often the only market-place for young new producers. Without Den Permanente even such a distinguished figure as the silversmith Inger Møller, honoured in 1966 with a one-man show at the Copenhagen Museum, might never have reached the national consciousness.

America is the land of extremes but even there the craft movement has failed to explode, making only quiet progress. Mrs Vanderbilt Webb, the philanthropic heiress, started a craft shop in 1931 near her home in Putnam County. She had been depressed by the rubbish in shops and appalled by the effect on craftsmen of the 1929 slump. This shop was her answer and it led her further than she knew. In 1940 she founded America House on Madison Avenue in New York as a centre of fine work there. By 1966 its year's turnover had risen to $340,000, and commercial realism extended to a retail mark up of 100 per cent and a commission of 30 per cent or 40 per cent on wholesale contracts. In 1943 she founded the American Craftsmen's Council which now has a modest 28,000 members. 1956 saw the culmination: the Museum of Contemporary Crafts was started near the Museum of Modern Art in New York, with America House now across the road. Ten years later, the tiny Museum West opened in San Francisco to take the best exhibits from New York. This was activity indeed, but on a humble scale for so vast a country. American craftsmen are often teachers at one of the few art schools like the Cranbrook Academy in Michigan, and seldom make their living with their hands. Henry

Shawah, the jeweler in Boston, is in this respect exceptional if not unique. American mass-production still wholly rules its own market-place, the richest in the world.

In Germany, help for the crafts first arose through the Deutscher Werkbund, always very close to industry, with such firms as Pott and Wilm actively participating in its counsels. Most of this type of activity is, however, characteristically official in its nature, by contrast for instance with England where it remains unofficial, or Italy where it does not exist at all. Each German state, such as, for example, Bavaria or Westphalia, has its own Handwerkskammer whose officers organize exhibitions, publications and educational grants in their regions. Most towns also have their own Innung or guild supported by contributions levied from all local firms. Regional government is much stronger in Germany than in any other country and these local organizations, between which there are continual exchanges, are therefore influential and important.

The pattern is repeated in most countries today: well-intentioned private societies try to keep alive techniques that are outdated and useless, to sell products that are unfashionable and therefore unsaleable, to perpetuate the myth that what is difficult to make must be intrinsically desirable. Some of these associations are, in fact, successfully humanizing society from within; but society has long ago rejected most of them. Missionary zeal is no substitute for commercial competence and it has proved extremely difficult throughout the world to achieve a happy compromise between the two. The World Craft Council, founded in 1964 by Mrs Webb, is still mostly American, but aims to strengthen diverse craftwork everywhere.

In England after the war the Crafts Centre of Great Britain was founded to stimulate skill. It concentrated at first on hand-made masterpieces selected by a committee, but in 1967 changed its policy, admitting all good work on the principle that it doesn't matter how things are made provided the end product is right, and having only one selector, who changes regularly, on the principle that committee decisions, on art subjects as elsewhere, are almost always boring and uninspired. No committee ever could decide what is a modern masterpiece. The fine new Crafts Centre opened in 1967 with a restaurant in picturesque Covent Garden, near the opera. Its aim is to encourage youth not by talking but by selling. The theory is that if craftsmen are rich, society respects them and then their recognition is won. The 1946 example of the Council of Industrial Design with its Design Centre in Haymarket, London, has since been emulated the world over, in Toronto and Amsterdam, in Darmstadt and Stockholm, in Brussels and Warsaw. Here is the home of mass-production, and in it the old craft industries play only a small part. The idea is not to sell directly, but to entice manufacturers to be more enterprising by exhibiting their new produce, giving it publicity, and therefore making it saleable by retailers.

Since they first arose in England about 1800, retail shops have proved indispensable. They are criticized by the public as being too expensive, by factories because they often double the price of factory goods, by designers because they suppress individual names, themselves claiming credit for all their stock, and by everyone because they are so slow to pay their bills. The role of the middle man is thankless. But town life without retail shops would be dull and drab, and most retailers, though they may not see themselves in such a light, provide a splendid battle ground. The retailer interprets public demand and harnesses factory production; he provides a choice. His mistakes may lead him to disaster, whereas with the craft societies mistakes lead only to regrets and a change of committee. And nothing sharpens the wits more powerfully than threatened disaster. Hence the difference in commercial acumen between the hazy do-gooders of the benevolent associations and the highly-charged businessmen of the shops.

Many retail firms run a small workshop of their own. Above Asprey's in Bond Street, London, is their fascinating factory where they make highly specialized leather and gold work, employing some forty men. Beaumont of Lyons, France, have perhaps half-a-dozen craftsmen on the ground floor in the main shopping street of this very expensive city, making fine jewels to customers' requirements. A. J. Smith of Aberdeen run on similar lines. Cartier of London have a jewelry factory on the premises and so have Garrard's, the Crown Jewelers, each employing two dozen craftsmen. Bolin, the Stockholm Crown Jewelers, adopt the more usual arrangement of employing their dozen craftsmen in an unfashionable building far removed from their shop. The convenience of manufacturing on one's own premises is often outweighed in a smart shopping area by the enormous rentals. Most of the bigger retail firms have some productive capacity, and many of them repair watches, but this accounts for only a tiny fraction of their turnover. They buy stock from far and wide, and a retailer's success depends very largely upon his knowing the best sources of supply.

Retailers cannot exist without advertising: it is necessary to keep their name continually before the public. The more up-to-date firms, particularly in the USA, have a very large advertising budget, but even so they are all negligible compared with the big mass-production industries: in 1967, De Beers, advertising diamonds, spent £150,000 a year in each of their larger markets, England, France and the USA, and £70,000 in each of their smaller markets, Japan, Italy and Sweden. A brewer with thousands of retail outlets may spend £1 million annually this way. But retail jewelers do not particularly like brand names – they like to print their own names on all the produce they sell.

With big factories one cannot generalize: some, particularly big cutlers like Pott of Solingen or WMF of Geislingen, are known to the public, and stamp their name on many of their products. Most retailers find this helps sales. Two great British manufacturers, Percy Adie of Adie Brothers the Birmingham silver factory and Major Gilbert Dennison the biggest British gold watch case producer, used to argue this point fifty years ago: Adie put other peoples' names on his products, that is the names of the retail shops and wholesale dealers to whom he sold, a scheme which certainly had short-term commercial convenience; Dennison refused to sell anything that did not bear his own name. Both factories did well under these fine leaders, but in the event the Dennison theory established a longer lasting reputation and the Dennison factory survived longer.

Even if there were silver factories big enough, it is doubtful if retailers would want their producers' names to be as well known as they are with watches or with clothes. A retailer's strength is his own reputation, and he is always frightened that if his suppliers become more famous than he, they will cut him out and sell direct to his private customers.

Most artist-craftsmen do just this – they sell to the general public: the essence of their appeal is the direct human contact between patron and maker which gives a unique pleasure to both. So retailers, if they buy from artist-craftsmen at all, usually buy a product which is made available exclusively to one retail firm.

The key to success is seen more and more clearly to lie in publicity: all of us like to be up to date, and it is in newspapers and on television that we can judge where others already are, and where we ourselves may be going. So a successful silversmith will advertise if he can afford it and try to get publicity in every possible way. There are brilliant salesmen, like Bernard Copping of William Comyns in London, who make their customers feel privileged to buy from them, but they are exceptional; usually the maker is too small to go very far in search of new business. He waits for trade to come to him: he encourages it with his regular team of commercial travellers touring the country. But the main bait he uses is successful participation in public events.

There are promotions all over the world commissioned by governments, art societies, and trade associations. Many such exhibitions have taken place in London since 1966: Swedish silver was shown at the Design Centre by the Swedish Design Society; Danish Jensen silver was at Goldsmiths' Hall, shown by Jensens and the Worshipful Company of Goldsmiths; German Friedrich Becker silver was also at Goldsmiths' Hall; Swiss watches and clocks were at the new Swiss Government Centre; a mixed bag of Italian work went to the Italian Institute for Foreign Trade; Finnish products were at the Campden Art Centre for the art festival there; the competition organized by Pforzheim for a gold necklace to celebrate its 200th anniversary as a jewelry centre was at Goldsmiths' Hall; modern British civic and university plate likewise took its bow there before going to Leeds Art Gallery during the music festival, under the name 'Public Treasure – new gold and silver for our towns and universities'. British fine silver was shown and sold at the new Crafts Centre of Great Britain in Covent Garden. Finally, of no commercial importance, was the Victorian Church Plate shown at Goldsmiths' Hall, called 'Copy or creation', the story of artists about 1900 throwing off the shackles they had imposed on themselves about 1840. Some of these shows were intended simply to introduce merchandise to buyers, others to persuade the general public that silver is an art not a business; all of them hoped to get publicity.

The older generation of shop managers still speak nostalgically of the easy times before 1914. There was not much competition, quality was often considered more important than price and the pace of trade was leisurely. Even the geographical layout of the trade was static. Then London was and always had been the centre of fine handwork: there were dozens of workshops scattered through Soho and the Clerkenwell Road, none of which probably employed more than fifty people. In Birmingham, a much larger centre for the production of lighter goods, particularly of cheap jewels and plated tableware, there were, besides Adies, perhaps three or four large factories, each with 400 employees – for instance, Ellis, who inherited and still used dies of the eighteenth-century factory pioneer Matthew Boulton, or Barker, who specialized in elegant period reproductions often using the old bright-cut engraving technique. Most famous were Elkingtons, the inventors in 1840 of electroplating, who catered for every market, small and large, even undertaking electrotype reproductions of treasures in the London museums. Sheffield, the third great British centre, was the home of the heavy contract industry: a new hotel or steamship line would get prices from two or three of the two dozen big factories mostly built around 1800 and mostly producing quantities of cutlery, such as Atkin Bros, for long the oldest surviving firm, who also used surviving eighteenth-century dies; Walker & Hall, who at one time inaccurately claimed the invention of electroplating; Mappin & Webb, later to become the most versatile of all, tracing their history back to Joseph Mappin an engraver in Fargate from 1797 to 1817; William Hutton who with Harrison's were leading platers and introduced electroplating into Sheffield; and Dixons themselves whose nostalgic factory name, Cornish Place, refers to their pioneering production of Cornish tin; each factory would tend to have its own special area of interest, like Dixons and Australia, or George Wostenholm who named their factory 'Washington Works' to record their American trade.

Many of the early factories had their own trademark, like the crossed arrows of Huttons, or the bugle of Dixons: one of the best was the symbol IXL chosen in 1745 by George Wostenholm himself when he started his factory. It meant simply 'I excel', and that was the spirit of the early producers. The majority of mass-produced silver today has rather sadly yielded to the power of the retailer and therefore bears the name of the middleman, not the originator.

One of the big continental centres for cutlery was Solingen, much smaller than Sheffield (in 1908 the number of workers from there was only about 10,000), but always its chief competitor. Ruskin, asked why he had established his small museum for workmen in Sheffield, replied:

In Cutlers' ironwork we have (in the town of Sheffield) at this epoch of our history, the best of its kind, done by English hands, unsurpassable, I presume, when the workman chooses to do all he knows, by that of any living nation.

It may be that Solingen was better mechanized than Sheffield – certainly it concentrated more on knives and less on silver. Then there was Ramscheid and the already huge silver factory at Geislingen or, in France, the cutlery town of Thiers. In America, many factories grew up in Providence, Rhode Island, and these attracted European craftsmen, particularly from Sheffield.

The two decades between the wars saw the rise of the mass market, and with it the devastating seasonal trade. Some factories would find themselves without work from January to April and did half their business in the month before Christmas. Steamships and hotels became at once more demanding and less discerning; Roberts & Belk of Sheffield, having spent many months producing the best possible cutlery for the maiden voyage of the *Queen Mary* in 1936, were amazed to receive a telegram from New York asking for an immediate repeat of half the order for delivery at Southampton a few days later. Thousands of pieces had disappeared. Time took on a new prominence. Price became all important and the older firms often chose to go out of business – as they are still choosing – rather than lower their old standards or achieve economic viability with new equipment. Cheap goods, however, as Henry Ford had shown, were vitally needed. Such firms as Richards cutlers or Viners silversmiths have not regretted heeding his message, even though the splendid words 'made in Sheffield' or 'made in England' thus became of only academic interest instead of meaning that the goods were of magnificent integrity. Many factories deteriorated and shrank, so that original ability tended to move to the small workshops.

Since 1945 some of the best factories have moved to the country, like Walker & Hall at

Bolsover: working conditions are much better, and there is no heavy industry, such as Sheffield steel or Gothenburg ship-building, competing for labour. Girls have always worked in Sheffield silver, chiefly as 'buffer-girls', polishing silver with dirty sand, for which operation they often dressed up in newspaper – as practical as it was picturesque. But girls now no longer do only the dirty work, they manipulate the advanced machinery and their attraction lies not only in the fact that they are cheaper than men; they are quicker too.

In every country the silver industry is facing a crisis: if it becomes profitable to dispense with handwork then the big factories will win through. At present though, tiny back-room out-workers, with no overheads, can still often undercut the giants and therefore continue to flourish. Gense of Sweden, probably the most brilliant factory in the world, was in difficulties because of Japanese competition. In England some of the best old works have ceased production – Adies, Walker & Hall, Suckling, Atkin, Harrison & Howson, Gladwyn – the list of those who have succumbed even since 1964 is long and depressing.

Summaries of actual prices and quantities produced mean nothing: the numbers are too big. In the Sheffield Museum is exhibited the 29th million Surform file, made in 1966, the first one having been finished as late as 1951. The success of this ingenious British invention by Firth Brown is spectacular, and illustrates the vast potential demand for an object that is universally needed. In 1966 the new Birmingham factory of Arthur Price was producing 2000 gross pieces of cutlery per week, that is a finished spoon every five seconds; 80 per cent of these were stainless steel, 20 per cent electroplated nickel silver and none solid silver. Viners of Sheffield considered their new stainless steel cutlery, 'Chelsea', designed by Gerald Benney in 1963, a world best-seller: they were making 3000 or 4000 dozen pieces a week of this pattern alone. It is estimated that in Sheffield, England, there are now only some 11,000 workers, including 225 firms averaging five people each, up to the biggest, Viners with 1100. But this is still a personal trade where the differences between firms are more significant than the similarities. The structure of the trade is a reflection of human nature, too diverse to analyse.

Both the biggest groups and the smallest men are now apprehensive. It is only the most original producers who move from strength to strength: the artist-craftsmen like Brian Asquith of Sheffield or Friedrich Becker of Düsseldorf whose asset is their individuality, or the makers of high precision goods like K. Weiss of London, box-makers, whose products cannot be paralleled by hand or copied by machine.

In silver, as in most fields of civilized activity since 1918, it has been very much easier to buy than to sell. People and factories are nearly all over-producing. Occasionally huge orders, such as the recent British government commission (for the British army and all government can-teens) for millions of pieces of the specially designed new David Mellor cutlery 'Minim', brighten the gloom. Normally trade is highly competitive. Contract orders to equip hotels or airlines are the most desirable because they bring steady employment over a long period, but they are at the same time the most fiercely fought for and therefore yield the least profit margins. Most firms spend much of their time satisfying the troublesome and fluctuating retail market, and for this a good name is essential.

An amazing development since the war, which started in America and is particularly strong in England, is the premium business. In essence it is a gift from a manufacturer to encourage his customers to buy more of his produce. Often the manufacturer will, for a limited period, simply give a free dose of his normal produce, for instance a small extra tube of toothpaste with the large tube for which the customers pay. Customers who produce half-a-dozen tabs from cereal packets, jam jars, or coffee tins, may be presented with articles that are useful in the home such as a tin-opener or a carving knife. They are persuaded that they are getting something for nothing. Certainly these gifts have proved an effective means of increasing turnover more cheaply and quickly than normal advertising methods in, for example, the rapidly expanding soap and detergent market.

The exciting refinement of what is by now an established marketing technique is that a manufacturer often now decides, instead of giving these presents, to sell them, and because the quantities are enormous, the market research impressive, and because there are no middle-men involved, the price, even when not subsidized by the manufacturer, is wonderfully low. Cutlers like Sippels of Sheffield have been in this field for years, but it is David Mappin who has pioneered these activities in England, with Arthur Price and John Mason the Sheffield cutlers; their company, David Mappin Holdings, now has as its art director Ken Lessons, a moral history in himself.

He studied as a painter in Sheffield, then as a silversmith in London, at the Royal College of Art. Unlike some dedicated creators, he left his art school determined to find a meeting point between what he himself liked and what the public wanted. He decided on cheap modern jewelry, and was one of the first serious modern designers in England to enter that field. His wife lent him £100 and with this as their joint capital they started producing light copper jewels which they sold for £1 to £3 a piece. He made them, she polished them and stuck on the pins in their tiny Soho workshop, and once a week they spent a day on a sales tour of hair-dressing shops and boutiques. By 1967, after only eight years, Lessons' turnover in cufflinks alone was about £100,000 and his total turnover about a quarter of a million pounds. His factories in Sheffield and Enfield produce about 6000 jewels weekly, ranging from hand-made and die-stamped gold and silver down to lightest pewter.

Lessons first became interested in premiums when he heard of the 100,000 Brillo knives made for David Mappin by John Mason; inspired, he designed the Oxo gravy boats which cost 30s each, the result of an investment of £2500 in tooling; this was the first and probably the only premium piece ever to receive approval from the Council of Industrial Design. Orthodox trade never welcomes radical change, and retailers and their allies, such as the Council, bracket premiums with unsaleable poor quality end stock.

The latest mammoth order is for $1\frac{1}{2}$ million stainless steel bowls. These are commissioned by Maxwell House, to be given free with their 8 oz. jar of coffee of which the bowl forms a lid. In a shop the bowls might cost as much as 17s 6d but Lessons' price is 1s $\frac{3}{4}$d; he bought steel where it is cheapest, in Pittsburgh and installed one of the three or four multi-slide presses in this country. The multi-slide press is an advanced British invention which shapes metal by several operations at once; the pieces are mounted on big rotating mandrells, mechanically polished on the outside, but so beautifully tooled on the inside that no polishing is needed. At the F M E works in Brimsdown, near Enfield, the bowls are hardly touched by hand. Lessons says, 'If anyone has to pick a piece up and put it down during the production run, something is wrong; it adds 3d to the cost.' He is now working on a toast rack at 10s, with six napkin rings, but the soup company concerned finds these objects 'rather up-stage' for the market. Twenty-thousand Players pewter tankards for sale at £1 will be more sensible: liquid has to be contained before it can be drunk, whereas toast can be eaten without a rack. The premium business, within its limitation of universal demand, may bring a new source of revenue to the silver industry and a new artistic distinction to the table of everyman.

118 The Reed & Barton factory or
    'Taunton Britannia Manufacturing
    Company' Mass. in 1830

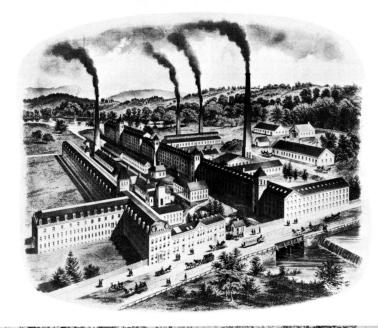

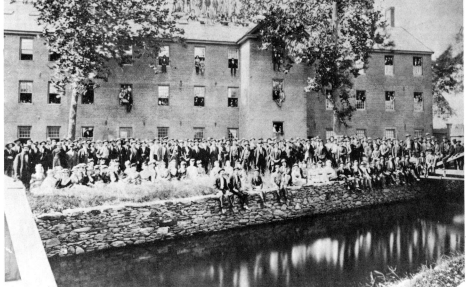

119 The Reed & Barton factory in 1880

120 A factory picnic at Reed & Barton in 1860

121 The Guldsmedsaktiebolaget factory in Stockholm, built in 1900, typical of many buildings all over Europe in which mass-production techniques are practised beside the old hand processes.

122 A relic of early display ideas when the frameworks were expected to be ostentatious in the same way as the objects inside. *Art nouveau* six-sided show case of glass, iron and nutwood.
   D Philippe Wolfers for an Antwerp jeweler's interior entirely by him 1896. Used at the Brussels World Fair 1897, the Paris *'Sources du XXme siècle'* exhibition 1960, the Essen *Jugendstil* exhibition 1963, and now at Hessisches Landesmuseum Darmstadt where it houses some of the Citroen collection of small *art nouveau* jewels.

123 Workshops and production usually mean mess. A fine modern exception, one of the smartest in the world, is David Mellor's at 1 Park Lane, Sheffield, architect Patrick Guest 1960.

124 The Bolin display at the Stockholm exhibition 1930. This was the world's first big show of modern design and had a great impact overseas, but sophisticated new ideas had still not penetrated the silver trade.

125 Typical Scandinavian shop: Bolin at Drottninggatan 15, Stockholm 1929 (see also 126)

122

123

124

125

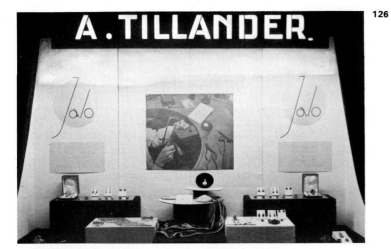

126 Tillander, Helsinki c. 1930

127 A smart modern interior. Old shops usually refuse to modernize for fear of losing their character and offending their established clientele. So it is usually only small new firms that try to achieve new architectural distinction.
D Paul Günther Hartkopf, artist-craftsman, manufacturer and retailer, Berliner Allee, Düsseldorf Architect Fritz Teschner, Düsseldorf

127

128

129

The three main window displays by Paul Günther Hartkopf during the British Week Düsseldorf when the city filled itself with British goods, 1964: windows
D Georges Pètremand, Frankfurt

128 Work by Gerald Benney, with photograph of Goldsmiths' Hall, London, home of the Worshipful Company of Goldsmiths of which he is a liveryman.

129 Work by John Donald with photograph of Goldsmiths' Hall, London.

130 Paul Hartkopf's own work with photograph of a new Düsseldorf skyscraper, the Rhein Rohr head office building.

130

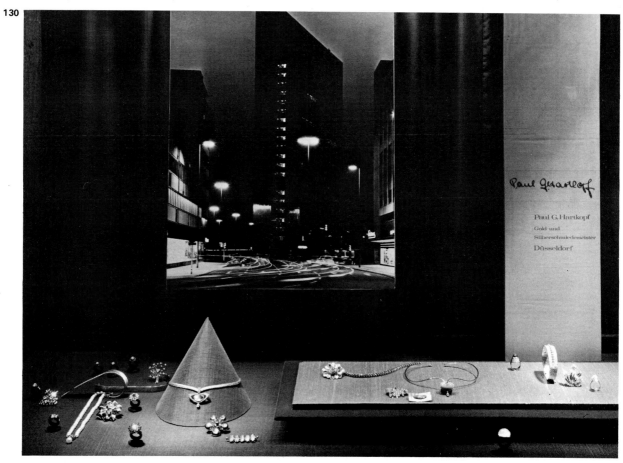

131

132

133

131 Andrew Grima's shop in Jermyn Street, London, exterior sculpture in slate and iron by Brian Kneale, aluminium door by Geoffrey Clarke 1967.

132/133 Andrew Grima's interior in which he sells Gerald Benney's silver and gold as well as his own jewels. The design gives prominence to a relatively small number of important objects, in contrast to most shops where there is so much on show that the best things may be overlooked. Architect Godfrey Grima, Andrew's brother.

134

134 The Worshipful Company of
Goldsmiths' stand at the British
Trade Fair, Tokyo 1965, interior
D Alan Irvine

135 The same, exterior
D Alan Irvine

136 Free-standing exhibition stands
made possible by quick modern
communications, have been an
important outlet for modern
display designers since 1945.
Eye-catching without to attract
visitors, and intimate within to
facilitate sales, their temporary

nature often inspires brilliance.
The Worshipful Company of
Goldsmiths' display at the British
Trade Fair in New York 1960,
showing some of the guild's
history as well as modern pieces
lent by manufacturers. Architect
Alan Irvine

135

136

137 A change of scale provides visual
    drama. A large perforated silver
    wall decoration at the international
    exhibition of modern jewelry
    1890–1961 at Goldsmiths' Hall,
    London, and in the touring
    Finlandia exhibition
    D Börje Rajalin
    M Kalevala Koru, Helsinki 1960

138 The best modern displays focus
    attention on the pieces, not their
    surroundings. Silver
    DM Sven Arne Gillgren
    M GAB at the Stockholm
    Design Centre, 1962

139 Advertising *c.* 1930. Crude and
    unpsychological by today's
    standards.

140 A stimulating clash of contrasting
    shapes. Candlesticks
    DM Sigurd Persson in his
    Hantverket, Stockholm
    exhibition 1964

138

139

UIFORCAT ORFÈVRE PARIS
RGENTERIE MODERNE DE JEAN PUIFORCAT

140

 141

142

143

Two commemorative stamps on sale in
1966. The artists might have preferred
help during their lives but official
gestures tend to be posthumous and
are therefore both safe and useless:

142 Georg Jensen

143 Josef Hoffmann (1870–1956)
'Founder of the Wiener
Werkstätte'

This year marks the
centenary of the birth
of Georg Jensen,
sculptor and silversmith,
whose genius has served
as a model and an inspiration
to the silversmith's art
all over the world.
He revived pride in modern
craftsmanship. The legacy
of his life's work
is a constant reminder
of the highest standards
of work and design

Graham Hughes

Art Director of The Worshipful Company of Goldsmiths, London

Glædelig jul og
godt nytår

Merry Christmas and
a happy New Year

Frohe Weihnacht und
ein gutes neues Jahr

Joyeux Noël et
bonne et hereuse année

GEØRG JENSEN SØLVSMEDIE

141 Christmas is becoming, amongst
other things, a pleasant occasion
for the exchange of trade greetings.
Front and inside of the Jensen
1966 Christmas card.

Stainless steel

A revolution made respectable

You can't have quality with mass-production. You don't want it because it lasts too long. So you substitute with styling, which is a commercial swindle intended to produce artificial obsolescence. Mass-production couldn't sell its goods next year unless it made what it sold this year look unfashionable a year from now. We have the whitest kitchens and the most shining bathrooms in the world. But in the lovely white kitchen the average American housewife can't produce a meal fit to eat, and the lovely shining bathroom is mostly a receptacle for deodorants, laxatives, sleeping pills, and the products of that confidence racket called the cosmetic industry. We make the finest packages in the world, Mr Marlowe. The stuff inside is mostly junk.

From *The Long Good-bye,* Raymond Chandler, 1953

In August 1913, the first experimental cast of chromium carbon steel was made by Harry Brearley at the Brown Firth Research Laboratories in Sheffield England. Their idea was to make better rifle barrels but, without knowing it, they had actually started a whole new industry. Stainless steel has extraordinary strength, and it can withstand heat and corrosion. It made the jet engine possible as well as modern surgical equipment, domestic sinks and washing machines, chemical, dairy and food processing equipment. Tableware and cutlery at first were clumsy and blunt, but the metal's indestructibility and its convenience as a heat insulator won through at last. Some designers today find cast stainless steel more beautiful even than silver, and most factories take the production problems in their stride.

The story begins in 1768, when J. E. Lehmann described a new mineral from Siberia, a form of lead chromate. In 1797, Vauquelin's analysis showed the lead united to the oxide of a new metal which he called chrom, from the Greek *chroma*, or colour. The first recorded use of chromium with iron was by Stodart and Faraday in 1820. Wood and Clark patented the addition of chromium to iron in 1872 and in 1892 Sir Robert Hadfield published his important experiments. Guillet and Monnartz developed the theory of the new metal. If Brearley did not invent it, he certainly launched it, persuading a Sheffield cutlery firm to make knives with it.

British and German firms, after 1918, pooled their knowledge. The British alloy of 18 per cent chromium with 8 per cent nickel, which proved better than the German of 20 per cent and 6 per cent, was accordingly launched in 1923 and ultimately became the world's basic standard. The problems of welding were solved. In 1926 the new alloy made possible the huge ICI synthetic ammonia plant at Billingham. Questions about corroding rivets and tempering blades, about work hardening and annealing, were asked and answered, often in the experimental factory set up for two years by Brown Firth to examine and exploit this British invention. The world stage was set in Sheffield to make tableware everlasting, but the first and only British firm then to take to it was not in Sheffield at all. It was J. & J. Wiggin of Walsall near Stafford.

In the 1890s shoe or belt buckles and horse harnesses were a big industry in Walsall. James Thomas Wiggin, grandfather of the present head of the firm, and his eldest son, James Enoch, started making buckles in a small converted stable next to his home in Revival Street, Bloxwich. By 1901, all six sons were in the business, now called J. & J. Wiggin Ltd. They brought to it their strong Christian outlook, symbolized in their own biblical names, Enoch, Hiram, Joseph,

Samuel, James, Noah and William. An old Salvation Army Mission Hall – known locally as the 'Old Hall' – was bought for £160, the combined family small savings, and converted into a factory. By 1904 there were over thirty people working there, and Wiggins became a private limited company.

William soon emerged as the leader of the firm, a position which he maintained till his death in 1957 at the age of eighty-two. His family now gratefully remember one of his first typical enterprises. In 1911 he went to South America on a tramp steamer to sell bits and stirrups: his total expenses were £100 and he took orders worth £16,000. One of his last and equally typical acts was the presentation of a new series of long-service awards to the company's employees. He set an adventurous and unusual business course, always accompanied by exceptional integrity in matters of production and of personnel.

Progress was steady. In 1913 the Wiggins bought a nearby factory for bridles and stirrups – V. Brodhurst of Eureka Works, High Street, Bloxwich, themselves founded in 1780, slowly coming to concentrate there on the drop-forging of pipe flanges for plumbing. In 1915 James Thomas Wiggin, the founder and chairman, laid the foundation stone for a new building, foundry, machine shop and warehouse. Munitions were succeeded in peacetime by roller skates and motor-car windscreens for Ford and Standard cars, but by 1920 William Wiggin decided to specialize in bathroom fittings, cleverly anticipating the new mass demand for hygiene. For two decades Wiggins were the largest manufacturers in the country of taps and plugs and chains, not to speak of stoppers and valves for hot water bottles and lifebelts. In 1925 they started the Oaks Manufacturing Company in a disused builder's yard, making toys and tin-openers, golf putters and cake trays, bathroom sponge and soap holders, all very cheap and mostly sold at Woolworths where the highest retail price was 6d. There was strength in this versatility. Even so, when the Old Hall was burnt down in 1928 the firm nearly foundered. Mechanical expertise was already the habit at Wiggins, but they were still typical of dozens of small family concerns in the Midlands of England.

It was in 1928 that they became unique. William Wiggin tried stainless steel. He had always wanted a metal for his bathroom accessories that would not corrode and did not need plating, and this last was it. 'Staybrite', the trademark used by Thomas Firth & Sons Ltd of Sheffield, was considered an industrial metal, and apart from knives which always seemed blunt, it had not yet found a domestic use. Wiggins were helped in their pioneering work of taming this tough material by Dr W. H. Hatfield, head of the Brown-Firth Research Laboratories. Soon they had made the first toast-rack, crude but functional, and as always well finished. In 1934 they exhibited the world's first stainless steel tea set. It was their main contribution to the popular spectacle 'Staybrite City', Firth Brown's big feature at the *Daily Mail* Ideal Home Exhibition in Olympia London, and it was the forerunner of the whole Wiggin series of pre-war tea and coffee services. The conservative silver factories in Sheffield refused to use 'Staybrite'; the general public, however, naturally put convenience before appearance and the new product was slowly accepted. But it was a very gradual change in the market. Many retailers, despite Wiggin's modest press advertising and exhibition promotions, would only use these strange objects on sale or return, declining to risk their money in such a rash purchase. The family were excited when they made their first sales in 1934 to Harrods and soon after to Queen Mary herself: stainless had not yet arrived in high society but it was at least being tried there.

Designs were at first all by the Wiggin family. The handles of the pots were nearly always of two strips of metal, the patented 'Staycool' handle which alone among mass-produced pots does not burn one's fingers. The surface was normally a bright, mirror finish, the shine on which is the criterion of excellence for so many factory products. Firth Brown thought a trained designer might bring further stimulus to the project and retained Harold Stabler, one of the most famous living silversmiths and industrial designers, to evolve a range of prototype ideas. Some of these, using sand-blasted or etched patterns, are in the collection at Goldsmiths' Hall, London; others were produced by Wiggins to whom Stabler was lent for the purpose. But they were put on sale only just before the 1939 war, and despite good publicity they were a commercial failure; they were very difficult and expensive to make, and the buyers all preferred more conservative shapes, often covered with hammer marks. Machine-made Staybrite must still pretend to be hand-made silver.

Not only the Wiggin pieces themselves but their advertisements too seem, in retrospect, almost quaint in their simplicity: an early catalogue of about 1938 reads:

> *It is truly said that they keep hot longer. Do you know that a fall of one degree in the heat of water when applied to tea greatly diminishes its flavour? Two or three degrees spoils it entirely. 'Staybrite' steel is a great resister of heat especially when highly polished as are these pots and jugs – and so the heat is preserved. The first few minutes makes all the difference. In British Hotels and Restaurants now, ladies prefer their tea and coffee served in 'Staybrite' pots. They ask for it to be so.*

Or, *A large percentage of 'single' orders received through stockists and agents are accompanied by personal letters with the Bride's address.*

These efforts, tentative but promising, failed to achieve their deserved triumph because of war.

Wiggins again turned to armaments, including secret devices for aircraft and submarine detection, leaving the consumer field open for the Scandinavian producers. In Sweden these remained at peace, in Norway and Denmark they were under German occupation, but in all three steel was more easily found than silver, and there was no patriotic call to change the factories or to close them down. Some had been introduced to the idea of steel for the table by Brown Firth who had been depressed by the poor British response to their initiative and who therefore turned to new markets, which they still largely supply. In doing so, they helped to create foreign competition and so stimulated England. The sophisticated, well-educated Northerners rose to the occasion. A soft satin finish was evolved which flattered the metal in its own right, instead of making it look like something else, and excellent names like Gense of Sweden, Tostrup of Norway, and Jensen of Denmark all identified themselves with stainless, often retaining magnificent designers like Folke Arström to lend it their distinction. By 1945 Wiggins were no longer alone: they were indeed rather behind their continental competitors in the vital matter of design. They were reluctant to adopt the fashionable 'dull' finish, partly because William Wiggin, now an old man, disliked it, and it was not till 1958 that their sales of 'dull' overtook those of the original 'bright' finish. Even in 1966 the bright finish still represents some 30 per cent of sales, and is popular in Norway and Denmark simply because nobody makes it there.

Reversion to peacetime work was slow: 1946 saw the manufacture of Fordson tractor cylinder-head studs and other parts, some plant being kept going during the fierce winter by tractors themselves! Leslie and Wilfred Wiggin did most of the designing for castings and chain and even for the machinery making them, as well as for tableware. In 1946, in line with government policy, agents were for the first time established in South Africa, Australia and New Zealand. Domestic sales were very disappointing but hotels, hospitals and caterers bought with increasing confidence, and Wiggins themselves therefore became increasingly aware of the need for structural precision.

In 1950 chain making was moved to nearby Cannock, in 1951 the company was renamed Wiggin Chains Limited, and by 1958 wholly automatic production was achieved, making Wiggins one of the world leaders in this field too. About 1950 there were 200 employees achieving a £200,000 turnover on tableware and general engineering, chains and castings accounting for a further 100 staff and £125,000 turnover; by 1955, despite a damaging universal shortage of the crucial raw material nickel in 1951–3, the combined sales reached £500,000, and tableware was moving ahead of the other products. There were changes in the financial structure: in 1960 the company was made public; in 1961 friendly arrangements were made for marketing with Hawker, Morris Ltd of Birmingham; in 1963 for cutlery with Harrison Fisher & Co. Ltd of Sheffield; the old Walsall iron foundry was turned into a polishing shop. In 1964, the company's diamond jubilee, the group sales reached £1,000,000 and in 1965 Old Hall Tableware Ltd alone exceeded this figure, the group reaching over £1½ million. There were 2000 retailers in Britain alone, and agents in 15 other countries. In 1967 growth continued with the purchase of Cheltenham Tool Co, makers of 'Lifespan' holloware. The struggle for stainless was decisively won.

Designers often worry that their sacred calling is being prostituted, that instead of improving the world their dedication is diverted to mere advertising, to styling, to change for the sake of change, to what the Americans call 'planned obsolescence'. Managers conversely often lament that designers are not practical, that many fine designs 'crash in harness' because they aren't right for the job, that advertising is more important than design, because it makes the difference between prosperity and poverty – 'There is nothing more dishonest in advertising than there is in people themselves' – even that good design does not pay. In fact, of course, there is room for change in most designs: the world is grateful for two versions of Leonardo's *Virgin of the Rocks*, and for three of Uccello's *Rout of San Romano*.

None of these philosophical superfluities disturbed the natural friendship between the Wiggins and Robert Welch. He was introduced to them in 1955 through the Royal College of Art where he studied silversmithing and wrote a thesis on the theory of stainless. His geometrical ideal suited the material, and his passion for precision suited the factory. His designs were not at first commercially successful, but he steadily gained experience and prestige as the firm's older, more curly products continued an asset to the balance sheet. In 1958, the splendid Welch-Wiggin association became known to the world when Welch's 'Campden' toast-rack and coffee set, named after the village of his workshop, won a Design Centre award, and Welch made his impact on the company accounts with a vast order given to him by the Orient Line to equip their advanced new liner *Oriana*. In 1962 came the second award from the Council of Industrial Design, for simple dishes. The judges had been unable to find any other range, British or foreign, offering 'this quality finish at this price'. A smart showroom was opened in Oxford Street, London, compensating for the firm's rather reserved attitude to exhibitions. In 1965 'Alveston' cutlery, called after Welch's Stratford-on-Avon home, won the third Design Centre award in seven years, and with it Welch's first substantial commercial success in the retail market. Good design had paid at last.

Welch's contribution was partly to bring beauty to the scene. But he also solved problems – and this is the difference between an artist and a designer. An artist may feel bored by the trivialities of the shop floor, he may not mind if one extra handling process adds 6d to the final cost. He may not even know how caterers manage to tear Wiggin handles from Wiggin bodies to which they are welded by the most modern techniques, leaving a hole in the body where the weld has been. To Welch, this sort of problem is part of the fascinating creative process, as important as, probably even a guiding factor in, deciding the final shape.

Leslie Wiggin, now the head of the firm, writes that he cannot remember Harold Stabler well himself 'except that he was a great perfectionist and his insistence on detail was probably a good experience for us'. When Robert Welch arrived on the scene twenty years later his 'insistence on detail quality was not completely unexpected'. Seldom do management and designer provide such exemplary mutual help.

Teapots are the Wiggin spearhead, but the cheapest Wiggin teapot is still about 57s. In densely populated areas, each person spends only about 6d or 1s on a Wiggin product. Even so, the factory cannot satisfy demand. More automatic machinery is needed, more streamlined production with less human unevenness. Welch's new 'Bistro' cutlery, the world's first stainless pattern with laminated Swedish wood handles, is so popular that new buildings are needed for it both in Sweden and in Walsall. Aesthetics and technique are here marching together, and a formidable pair they make.

144
'Richmond' stainless steel teaset with patented 'staycool' handles, still one of the very few handles in production which do not burn the user's hands. The fashionable square and step shapes were unsuited to metalwork
D Harold Stabler
M J. & J. Wiggin, *c.* 1936

144

145 Prototypes for mass-produced
    etched stainless steel.
    DM Harold Stabler
    M Brown Firth, *c.* 1930
    O Worshipful Company of
    Goldsmiths

146 'Alveston' stainless steel cutlery
    Design Centre Award 1965.
    D Robert Welch
    M Old Hall Tableware Ltd, 1963

147 Nutcrackers, stainless steel.
    D Robert Welch
    M J. & J. Wiggin Ltd

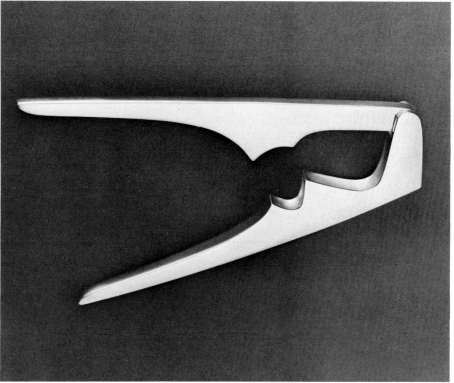

148 'Alveston' stainless steel teaset.
   D Robert Welch
   M Old Hall Tableware Ltd, 1962.

149 'Avon' stainless steel coffee set.
   Handles made by the lost wax
   process.
   D Robert Welch
   M Old Hall Tableware Ltd, 1967
   O 146–9 Worshipful Company
   of Goldsmiths.

The Pioneers

All my life I have been trying to make beautiful things and now that I can make them nobody wants them. Only my extinction can make them valuable.

Said in the last year of his life by William Frend de Morgan (1839–1907), and quoted with a fine article about him in *Apollo* magazine, published not when it would help him but in 1967. He was perhaps the most distinguished British potter of the nineteenth century, lifelong friend of William Morris and Edward Burne-Jones

If fashion decides what people want, it is designers who decide how they shall get it. Designers create the range of ideas from which the public make their choice, and many of the ideas are of course either never chosen, or chosen so sporadically that in retrospect they do not matter. It may seem silly to present so small a number as four designers as the pioneers of twentieth-century silversmithing, but without these the pattern of development in gold and silver would have been different. There are, of course, many other designers whose work may sometimes have been more beautiful, sometimes more lucrative, sometimes more successful, but the four now examined are outstanding. They have each achieved the impossible: a blend of novelty with mature style, of commercial adventure into the unknown with long-term commercial success.

Georg Jensen of Copenhagen is unique because he was both a man and a firm, but it was his own expansive personality that made necessary this remarkable growth. He stamped his personal *art nouveau* on the consciousness of the world, the only *art nouveau* designer in any field who interpreted the style sensibly and, therefore, the only one whose work stayed in production. Meinrad Burch-Korrodi of Zürich is a fine jeweler. He has made one of the best collections of precious stones, and runs one of the smartest retail shops. But his unique contribution has been enamels; he has sold every variety of enamelled gold and silver almost everywhere, reminding the world that gold and silver can again be coloured as they once were at Verdun or Limoges. Sigurd Persson of Stockholm has brought fun to the functional movement, contriving to give personality to his very austere designs, showing that plain surfaces need not be dull. Last and youngest of the group is Gerald Benney of London, probably the only silversmith in the world to specialize in hand-made commissions; he has done for silversmithing what architects do for building, persuading clients that tailor-made is better than ready-made. Commemorative silver sounds grim. Benney has made it at once a result of and a cause for celebration.

Jensen was born in 1866 at Raadvad near Lyngby, outside Copenhagen. His father was a cooper; as the boy liked handicrafts, he was bound a goldsmith apprentice under several masters, one of them Mogens Ballin whom he joined for three years at the age of fourteen. He became a journeyman in 1884, studied at the technical school and the academy of art where he did a bust of his father, and emerged fully trained in 1892. His first notable work, *Harvester*, was exhibited at Charlottenborg that year, and attracted such attention as to suggest that he had already arrived. Actually he had failed. He couldn't make a living. Nevertheless he forgot silversmithing and decided he was a sculptor. He earned good notices at his 1898 Academy exhibition, and in 1900 in the Danish pavilion at the Paris World Fair he won a *mention honorable* for the ceramics he exhibited with Joachim Petersen: some of their pieces were bought from the exhibition by the Copenhagen Museum. Also in 1900 he was awarded the Academy travelling prize, and studied in France and Italy. He seems then to have realized that many great artists were designing utility objects, and that handicrafts need not be ignoble; this was a decisive revelation for him.

About 1903 he resumed his old craft, making silver for Mogens Ballin and, for the first time, exhibiting silver jewels. Perhaps he suspected that, having foundered as a sculptor, his pottery too was doomed, perhaps family financial worries made him feel he must try every commercial outlet: in 1897 he had, for the first time, become a widower, with two small children. He soon broke with Petersen, rejoined his teacher, and became foreman; he opened on his own as silver-smith and jeweler in 1904 with just one apprentice and a tiny showcase in the entrance gateway at 36 Bredgade in Copenhagen. There his early products sold, even though today they seem heavy and homely. Amazingly, however, it was in that same year that he designed 'Antique', one of the most distinguished cutlery patterns of the century, still in quantity production today. At last, at the age of thirty-eight, he was started.

His first modest foreign exhibition was at Hagen in 1905 where the Director of the Volkwang Museum recognized Jensen's gifts, knew that no German goldsmith shared them, and urged him to open a shop in Berlin. Thus in 1908 Jensen had a permanent enterprise abroad, although his further plans were stopped by the war. His workshop expanded continuously. In it he trained some of the leaders of the next generation, like Mrs Just Andersen, Kay Bojesen and Inger Møller. With Denmark, his best market was Sweden where he did well at the 1914 Malmö Baltic Exhibition. His chief Swedish supporter was the famous art dealer Nils Wendel who influenced the Royal Family patronage, but surprisingly the Jensen shop in Stockholm was not open until 1930, ten years after Paris in 1919, and London in 1920.

Jensen was intensely alive as a person, and evidently believed that vitality was in itself an asset, even though its results might not be of immediate financial value to him. When in 1930 his old friend and colleague Harald Nielsen showed him his models for the new cutlery pattern 'Pyramid', Jensen said, 'It's neat, but it won't sell.' Nevertheless, he decided to produce it, respecting Nielsen's faith in his own work. Nielsen now tells with a chuckle how 'Pyramid' has become one of the two or three best sellers in the whole history of the firm. Jensen's reputation grew steadily because he continuously patronized new designs, never resting on his laurels, never showing a personal prejudice in favour of his own work and against that of his colleagues. He cannot have foreseen the problem he would create: so many of the designs had the combination of vitality and soundness that now earns them the epithet 'good', that there is still a steady demand for what Jensen assumed would die a natural death. By 1960 there were some fourteen thousand Jensen patterns in production, the number now being reduced to perhaps ten thousand. Jensen would have been bewildered by the scale of his success and by the storage problem it involved!

Jensen was as prolific matrimonially as he was with silver: he married four times, each wife dying before the next marriage, and each producing children; his family connections are now, therefore, widespread. His wives were obviously important to his rumbustious emotional life. His third wife (her sister married the designer Gundorph Albertus), whom he married in 1907, was perhaps the most important from a worldly point of view – it was the only marriage that had any financial advantage; his wife's brother, a lawyer, named Møller, country gentleman and amateur cellist, interested himself in the growing business, and about 1916 even lent such modest support as he could afford. In 1920 the Møller family and the Hostrup-Pedersens, who had recently come to be jointly responsible for the financial soundness of the firm, divided. The Hostrup-Pedersens took over the factory and the Møllers, friends of the Wendels, took the retail shop in Copenhagen called Jensen & Wendel, now managed by Jorgen Møller.

It is a cliché of modern business that no luxury consumer goods firm can thrive unless it succeeds in the USA. Fifty years ago this was not so obvious. But Jensens even then were ambitious for world fame: in 1915 they sent William Arup to the San Francisco World Fair with a collection, and there they secured their decisive American success. William Randolph Hearst, characteristically, having enjoyed the exhibit, bought it all, thus making Jensen the only smart name for table silver. All Hearst's many friends and competitors felt they must emulate him, probably the biggest magnate in the country, and certainly the most famous. Just Lunning, the family friend who had sold Jensen's goods in his gift shop in Odense without much success, started a new life in New York where he opened the Jensen shop in 1920, quite independent of the parent firm but in friendly association with it.

Most of the Jensen catalogues reproduce a photograph of the man himself in his artist's smock, and rightly emphasise his achievements as artist more than financier. In 1923, for instance, the Copenhagen factory issued a picture booklet with an impressive list of awards:

Grand Prix, San Francisco, 1915
Grand Prix, Rio de Janeiro, 1923
Médaille d'Or, Brussel, 1910
Diplôme d'Honneur, Gent, 1913

Pieces in national museums are then reproduced and, as not all of them had a progressive buying policy, Jensen must already have been considered an extraordinary phenomenon: 'Kunstindustrimuseet, København; Nationalmuseet, Stockholm; Rohska Museet, Göteborg; Vestlandske Kunstmuseum, Bergen; Kunstgewerbemuseum, Köln; Museum Folkwang, Hagen; Musée d'Art et Histoire, Genève; Musée des Arts Décoratifs, Louvre, Paris; Metropolitan Museum of Art, New York; Newark Museum, Newark, NJ; Detroit Institute, Detroit.' Retail agents are given for Copenhagen, London, Paris, Amsterdam, Detroit (Michigan), Glasgow, Lyons, Marseilles, Stockholm, The Hague, New York, Bradford, Edinburgh, Liverpool, Manchester, Oxford, Southport. In 1923 Jensen also won the Grand Prix at Barcelona, and almost every year since then has brought a major world award. It is important that Jensen himself formed and inspired the fine team he left behind him; but his unique and personal achievement was to prove to the world that modern silver could be art.

Meinrad Burch-Korrodi of Zürich had an easier start than Jensen, but he chose a more original course and pursued it with more certainty. Born in Giswil, Obwalden, in 1897, he probably could have adopted any career he wanted. At school he could not stop drawing, and his friends felt sure he would become a painter or sculptor. His father, however, who was a poor grocer, could not believe that Meinrad would be a real artist and insisted on his taking a humble goldsmith's apprenticeship in Lucerne. Then he went to the famous London Central School of Arts and Crafts which had been founded by William Morris's friend Lethaby to encourage fine craftsmanship. He says of his training there that it was 'technically of a very high standard, but there was never a discussion about a new idea and this was especially valuable for me. The silversmiths copied Sheffield silver based on catalogue pictures.' He learned to use his hands efficiently and to respect ancient skill.

Next he worked mounting very expensive jewelry for the Bond Street shops. He thought of London in 1922 as a sort of Ellis Island: all foreigners seemed to him to be treated as Hottentots; Meinrad was terrified of policemen, hungry and out of work. He jumped for joy on the pavement of Hatton Garden when he was at last offered 1s an hour by an old Jewish maker of platinum and diamond jewels named Wachkirsch. Soon he was put up to 1s 6d, and he remembers making a bracelet with 650 diamonds. By 1924 he was ready for a change. In New York he got a job with German Jews, Davidson and Schwab, opposite the public library on 40th Street. After a spell in Paris, he returned to start his own workshop in Lucerne in 1925. In 1932 he moved to Zürich, a bigger, richer, and more cosmopolitan centre, where he hoped he would find more scope. He was right.

He started making chalices about 1926 and used to get them enamelled in Geneva by two old German engraver/enamellers called Arnold and Steinwachs. After half-a-dozen pieces had been finished in this way, Burch became intoxicated by the technique and began to find his own distinctive art. For a few years he employed a Norwegian enameller, named Berger, who had been teaching in some German schools. Production increased, and, as always happens with a strong designer in charge, style therefore became more confident. Burch sold to every sort of church, Protestant as well as Roman Catholic: one of his best patrons was the Lutheran Valparaiso University, outside Chicago, who recently made a colour film about his work. After thirty years, in the 1960s, he was producing four hundred chalices annually with some eighty in the workshop at any one time, all begun by machine, each finished by hand so as to be a unique statement. He employed no more than twenty craftsmen, but sold to almost every country except Russia, China and Italy. With an annual turnover today of some two million Swiss Francs, and a feeling that he has never betrayed his conscience in the cause of profit, Burch may reasonably claim an astonishing impact on Christian silver throughout the world.

Like many successful designers he made his reputation by taking part in every possible exhibition; but he avoided the common mistake of confusing reputation with prosperity. It is useless to be famous if one is starving. So his art activities were very sensibly based on a sound if modest turnover from perfectly made silver, jewels and enamel. His work is a splendid representation of the Swiss passion for precision.

He took part in many Swiss arts and crafts society shows. He also exhibited in New York at the 1928–9 international exhibition of art and applied art; in Dresden at the 1929 international exhibition of Christian art; again in New York in 1930 at the Metropolitan Museum in the third international exhibition of contemporary Christian art; in Geneva in 1931 at the national exhibition of art and in Rome in 1934 at the international exhibition of Christian art.

By 1936, when he won the Grand Prix at the Milan Triennale, he was widely recognized for his exceptional integrity and versatility. It is interesting to consider the public reaction to his exhibition work: some exhibitions attracted hundreds of thousands of visitors, some, like the Brussels World Fair, millions: hardly any of them yielded specific orders: nearly all of them caused controversy and discussion both in print and in private. Many 'exhibition pieces' are intended to cause a sensation, which may be no bad thing – the main purpose of silver and jewelry has always been to give visual pleasure; but for Burch, exhibition goods are as restrained as his normal shop stock. It is gratifying that critical opinion nevertheless recognized him so fully and so soon.

He was at the 1937 Paris international exhibition of applied art; at the Zürich 1939 Swiss national exhibition; in 1951 the German Goldsmiths' Society gave him their ring of honour – a pleasing personal gesture, the ring made by one member for the Society to give to another; he was at the Baden Baden 1963 showing of Swiss modern art. He won a gold medal in the Vienna 1954 display of Christian art and another gold medal at the 1958 Munich handicrafts fair, and he was well shown at the 1958 Brussels World Fair.

Throughout his career, Burch has maintained a fine variety: some designers are tempted by success either to specialize in some narrow field of activity which tends to make them artistically sterile, or to expand their business so much that they lose artistic control. Burch has done neither. He has always enjoyed silversmithing as much as jewelry, coloured enamel as much as coloured precious stones, window display as much as meeting customers. In 1966 he decided to discontinue silversmithing, but he still allows himself to be persuaded by an interesting opportunity. Today, in his small modern shop in Bahnhofstrasse, Zürich's smart shopping street, one may find three or four assistants; Burch himself is likely either to be in the workshop a few doors away in the city, with his two dozen craftsmen, discussing the intricacies of enamel texture, or he may be at the back of the showroom crouching over his drawers full of rare stones. He has wide interests, but perhaps his enjoyment of colour is his special pleasure, his particular contribution to skill.

Other silversmiths and jewelers have attained a comparable standard of austerity; his unique achievement is the revival of church art. His methods have not been a new expression of Christian passion but quite the reverse. His sacred vessels are always cool and well organized, and he uses enamel not only because he likes it, but also because, he says, he tries to 'improve technique and durability'. A service covered by enamel is always clean. It does not oxidize and does not show finger-prints. His sheet metal sculpture is also original – an art form in general almost forgotten. He brought dignity to his plate and grace to his sculpture, in contrast to many modern artists whose religious work is so uncouth that churches will not buy it.

In 1953 the Swiss Guild of St Luke exhibited Burch's work in Lugano at the third international Liturgical Conference, and then honoured him by publishing a book about him; its introduction finishes with the words 'may this book prove to be the foundation of the new Christian art'.

Thomas Telford, the English engineer, helped to make Sweden accessible by designing for it the Göta Canal in 1808–32; but for most Englishmen this led only to cold northern wastes. It was Stockholm town hall, built by Ragnar Östberg between 1911 and 1923, which made Sweden suddenly desirable. Here was one of the first ceremonial modern buildings. More significant, here was a message: advanced architects could and should use teams of humble artist-craftsmen. But it was not until 1930 with the Stockholm Exhibition that the rest of Europe suddenly woke up. Scandinavian modern design was not the same thing at all as Stockholm town hall, it was not even another development of hand craftsmanship. It was something totally unexpected: a whole new style of applied art resulting from widespread use of machinery where machines had never been before.

Fifty years ago machines in the old industrial countries, England, Germany and France, were mostly still producing old designs, and, mass demand always tending to be conservative, the old designs continued to be lucrative. In the north, by contrast, style was inspired not by what already existed, but by production convenience. There was very little historic tradition to draw upon – many of the leading Swedish and Danish court architects and designers had in fact been foreigners, and many of the smarter home interiors had simply been imported. Scandinavian designers had the rare good luck to start from nothing.

Compared both with Jensen and with Burch-Korrodi, Persson's scale of operations may seem modest. He has a workshop with only four craftsmen, no permanent retail outlet, and no great financial fortune. But his reputation is undeniable. Almost every Swedish tourist pamphlet mentions him, he is represented in many of the world's best museums and his 1966 one-man show in Stockholm filled four large rooms without including any of his older work. Of all the designers groping towards an ideal life in modern silver, he is the epitome.

It is not so much that he is exceptional, but that he is typical of a whole generation and outlook; applied art, he seems to say, should be sensible; the hard finish of machine work has real validity, even in the field of hand-made luxury goods.

He was born in 1914 in Hälsingborg, and received his first craft instruction from his father, F. S. Persson, also a silversmith, finishing his apprentice's trial piece in 1937. He went to the Akademie für Angewandte Kunst, Munich, from 1937–9 and to the Konstfackskolan, Stockholm in 1942, receiving his Master's certificate there in 1943. He opened his own workshop, and went through the usual agonies of trying to attract attention in an apathetic world. At that time he was artist-craftsman, not industrial designer, and he, like all such beginners, was dependent on the good will of individual men and women who bothered to come to him to commission and buy his exceptional work, rather than buying dull ready-mades over the shop counter. He travelled occasionally and extensively in Europe, but it was seven years before he had his first one-man show in Stockholm at the small shop Waldenströms Konsthandel. In Stockholm, as everywhere else in the world, small freelance producers need a combination of courage and brilliance, of craftsmanship and salesmanship, which is itself rare, but to which they have to add for their early years, indispensable and unexciting persistence.

Persson became convinced that he must give his energetic support to every possible exhibition, and he also developed an unexpected talent for exotic jewelry. He realized that exhibitions meant publicity even if no financial gain; and he sensed, being married himself, that every girl wants a jewel, so the market for jewelry was likely to expand as girls became richer. By contrast, the scale of private living was contracting and with it perhaps the demand for good silver.

It is surprising how few silversmiths can make convincing jewels, and it is damaging to both crafts that their producers tend to be so specialized. Not so Persson. But even with his silver and his jewels, he would hardly have succeeded without his industrial design. He worked on cooking equipment, stoves and kettles, for the big kitchenware business, Kockums; he designed cutlery for the Swedish co-operative society (KF or Kooperativa Förbundet), and heating controllers for Tour Agenturer, even decorations for buildings in concrete and marble. Each long-term commitment to a big factory yields a regular salary of inestimable value to an independent designer, whose occasional freelance commissions otherwise fluctuate in a manner alternately depressing and exciting. Some of Persson's industrial work earns him royalty commissions, some of it is done for annual fees, but all of it gives him stability.

In 1959 came his great turning point, when he won the competition for flight restaurant equipment for the Scandinavian Airline System (SAS), whose cutlery he supervised at Kooperativa Förbundet, and plastic holloware in Norway at Norplasta, Trondheim. These successes were followed by many varied commissions, even the designing of symbols, cap badges and insignia in brass for the Swedish army, and insignia and graphics for Umea University.

The Nordiska Company (NK), the huge Stockholm department store, have enough resources to take a risk; under Astrid Sampe their Art Director, herself famous in fabrics, they are a continuously stimulating influence. Their silver and jewelry department is probably the most varied in Sweden. Of unique value is their brave practice of buying nearly all the exhibits in any small exhibition they stage in the department, thus removing what would be an intolerable burden for an unknown craftsman. Persson was lucky enough in 1960 to stage there a show of '77 Rings', which sold out, and was then shown in Liberty's, London. In 1963 he showed equally remarkable bracelets, and in 1965, necklaces so large that the wearer hardly needed clothes.

Persson's other exhibitions are impressive: each one paralysed hundreds, if not thousands of pounds worth of his stock, but together they made him famous: he had one-man shows in 1956 Havana, Cuba; 1957 Nordenfjeldske Kunstindustrimuseum, Trondheim; 1959 Kunstindustrimuseum, Oslo; 1961 'Sigurd Persson Design', at Malmö Museum; 1963 'Silvery Candlesticks', Hantverket, Stockholm; 1964 'The Eloquent Jewels of Sigurd Persson' in Georg Jensen, New York; 1966, one-man show in Hantverket, Stockholm.

His awards show the admiration of the world's aesthetes: 1951 Diploma d'Onnore, Milan Triennale; 1954 Silver Medal Milan Triennale; 1955 The golden ring of honour of the German Goldsmiths' Society; 1957 Silver Medal Milan Triennale; 1958 Gregor Paulsson Prize; 1960 Silver Medal Milan Triennale; 1961 Swedish Form – Good Form prizes; 1965 Gold Medal at Ljubljana Triennale, and the Culture Prize of the city of Hälsingborg. In 1963 he was made a member of the culture board of the Swedish Ministry of Education, and a Design Associate of AID, and in 1965 an Associate of the Worshipful Company of Goldsmiths in London.

Arthur Hald of the Swedish Society of Design has described him as 'craftsman, designer, philosopher'; Dag Widman of the Stockholm National Museum describes his jewels as 'sculpture for the hand'. Persson supported the slogan 'comelier commonplace commodities', but he also pioneered the use of expensive and elaborate enamel, which he himself had learned as a lay brother in the Monastery of St Martin at Ligugé in France. The strength of Persson lies in his diversity. At 52 he has already made a notable contribution to most fields of Scandinavian art life.

Every art student thinks he can design an original teapot, and most can. But to keep up a continuous succession of new designs for such a utilitarian object is not so easy. Gerald Benney is one of the few successful designer/craftsmen in Europe whose main effort and income are both related to handwork, not to industrial design. He not only wants each successive teapot to be a unique personal statement and hates duplication, but, unlike many craftsmen, he really enjoys the creative process. It is a quality of uninhibited pleasure which makes even his plainer domestic silver so distinctive.

But it is modern ceremonial, rather than domestic work, that shows his style most clearly. He has produced an amazing succession of provocative ideas, for clients as diverse as the Royal Perth Yacht Club, Western Australia; London University, England; Metallgesellschaft, a steel company in Frankfurt; the Governor General of Tasmania; Meister of Zürich, Wako of Tokyo, and Hardys of Sydney, the leading retail firms. It is, unfortunately, true that private commissions for large works of art have been a casualty of the welfare state. Few people now live in such style as to accommodate in the normal routine at home, for instance, a great new silver centrepiece or a golden cigarette box for the table. But, as private life becomes less flamboyant, public bodies feel the need to establish their identity, and they often do it by means of ceremonial art. They commission a design from one of the daring young men who are now changing the meaning of modern British silver. Benney is foremost among these, and his larger pieces are ideally suited to this stimulating function: to decorate, to discuss, to give pleasure, above all to make individual what might otherwise be anonymous.

To many people, the words modern and plain are synonymous; only plain shapes do justice to a beautiful underlying form, so the argument runs, and decoration just conceals the lack of a basic, fundamental idea. But Benney is different. He uses exotic embellishments, such as crocodile skin and ivory, with the same assurance that he brings to the placing of hammer marks on a flat silver surface. He does not fit into any category because his smallest pieces are as magnificent as his most elaborate. He is, in fact, an extraordinary combination of vital imagination and sensitive craftsmanship. His exuberant fantasy is a tonic in the cold clinical atmosphere of modern functional design.

He was born in Hull in 1930. He studied at Brighton College of Art where his father was principal; then under Dunstan Pruden, who introduced him to the intricacies of church silver; in the army 1948–50 he painted some decorations in Aldershot Military Museum. Finally he went to the Royal College of Art, where he included furniture and mechanics among his other interests. Indeed, mechanics and speed were a minor obsession of his youth, and he expressed it on the race track: he has owned twenty-nine motor-cars, and twenty-two motor-cycles, fifty-one vehicles in four years. While still a student he helped to make the 7-foot dragon, designed by John Skeaping, which now surmounts the Dutch church in Austin Friars in the City of London, astonishing proof of his early manual dexterity.

In 1955 he set up his own workshop in London where he now employs six assistants. Progress was fast and sure. He took on a new man almost every year and made a small but steady profit of £500 or £1000, supported by his Viners income. But the financial strain of expansion is frightening: the difference between success and failure is a razor's edge. Because Benney is a good businessman, and because he infects his clients with his own love of the materials ham- mered gold and silver, he survives. Between 1964 and 1966 his costs rose by 100 per cent. There are few people succeeding like him anywhere in the world. This is mainly because Benney can make a pocket watch in two months, as he has just done for an American friend: he has almost unique ability. But the artist-craftsman's life in 1967 is always more a challenge than a routine. In 1960 he designed and made, in the course of one week, a large jug for the Worshipful Company of Goldsmiths, of which he is a liveryman. Their film, *A Place for Gold*, shows this water jug at various stages in its construction; but it could not convey the degree of energy needed to make such a piece so quickly. He repeated the feat in 1967 when he designed and, with his col- leagues, made five very large silver dishes in only three and a half weeks – gifts from the British government to newly independent islands in the West Indies, Santa Lucia, Antigua and others. The prime characteristic of many devoted craftsmen is patience. For Benney it is the restless striving of a real creator. His rapid success is all the more surprising in the context of the rather subdued role which most craft workshops play in modern industrial society. He has made plate for many leading companies, universities, and other institutions. Public patrons have backed him, and he has brought glamour to them both now and for the future.

Since 1957 he has been the consultant designer for Viners, the Sheffield silver and cutlery factory, where he spends perhaps one day each month. His 'Chelsea' cutlery pattern, in stainless steel, is probably at present the world's best selling design. 'Studio' was certainly the first British design to be entirely machine-made, and the first stainless steel anywhere to be deco- rated in high relief. He has worked for other industries, including Ridgeway Potteries and Glacier Metals. But silver is his main concern. He lives in Beenham House, a Georgian mansion outside Reading, Berkshire, to which he has moved half his workshop from London, with three of his nine craftsmen. He divides his time between there and London, pursuing his highly personal aim, a dramatic injection of youthful originality into silversmithing.

The small county borough of Reading contains Benney's biggest monument to date, a superb new civic collection designed and made by him for local firms to give to the town. Great Britain, more than any other country, is the home of fine public collections of plate, belonging to towns and guilds, to companies and churches, to universities and regiments. Here is evidence that silversmithing is one of the arts at which the British have really excelled. But often these public groups are an interesting muddle rather than a coherent scheme. Reading has now been pushed by Benney towards two unique distinctions: it has rallied private local patriotism, from laundries to police clubs, from breweries to biscuit-makers, to commission a magnificent series of new designs; and it has, with great courage and enterprise, placed all its eggs in one brilliant basket, thus achieving a civic symbol of fine consistency and strength.

Corporations are by nature cautious. Reading has proved the dazzling exception. It has moved with speed and imagination; here is the ideal art patron, spending contributed funds with the same dash as private citizens were once able to afford. Benney is Reading's goldsmith but he is also the world's, and he is still only thirty-six. He is one of a small group of young creators who have first made London capital of the art world and then persuaded us all that precious modern silver is not a wild luxury but a necessary part of civilized life. Benney's noble silver will bring credit to the name of Reading for centuries to come, and has already done so in the Worshipful Company of Goldsmiths' exhibitions in many countries. Let us hope that his example will inspire other towns to embellish their official life, and to put their faith in youth.

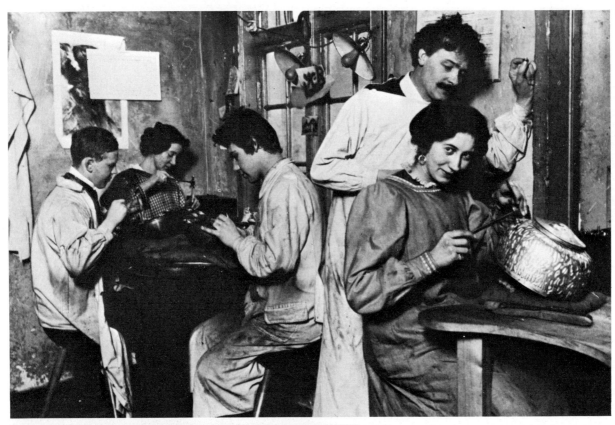

151

152

150 Georg Jensen's workshop in
Bredgade Copenhagen 1906.
(Such primitive equipment often
accompanies high ideals.) The
lady on the right became Mrs
Just Andersen, next is Jensen,
second from left is Inger Møller,
and on the left Kay Bojesen

151 Still a popular production pattern.
DM Georg Jensen, Copenhagen
1904, the first silver cutlery he
ever designed

152 Silver teapot, ht 105 mm.
DM Georg Jensen 1905, his
first piece of holloware
O Kunstindustrimuseum,
Copenhagen; National
Museum, Stockholm;
National Gallery, Melbourne;
Joslyn Memorial Museum,
Omaha, USA

153 Silver coffee set, ht 170 mm.
DM Georg Jensen 1906

154 Silver tea urn, ht 310 mm.
DM Georg Jensen 1918

155 Bowl
DM Georg Jensen 1918

156 Candelabrum, ht 265 mm.
DM Georg Jensen 1920
O Danish royal collection

153

154

155

156

157

Meinrad Burch-Korrodi claims that his designs have hardly changed in style throughout his life. Certainly his late work differs little from his early – for instance his figure of Christ for Sachseln Obwalden near Lucerne (161) which he first made in 1934. When he remade it in 1967, the later technique is more sensitive but the imaginative concept is the same.

157 Chalice
   DM Meinrad Burch-Korrodi,
      Zürich c. 1964

158 Chalice
   DM Burch-Korrodi 1965

159 Chalice with apostles
   DM Burch-Korrodi 1965

158

160

160 St Christopher, 15 in.
   DM Burch-Korrodi 1965

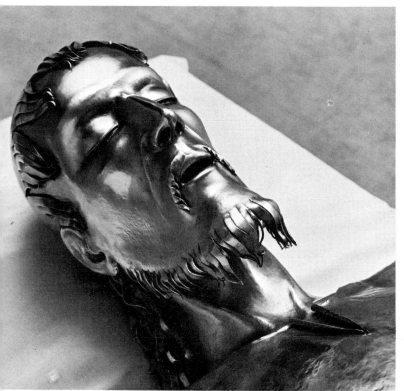

161

162

161 Recumbent Christ. Burch's early
work like this was more
sentimental than his later
DM Burch-Korrodi 1934
O Sachseln Church, near
Lucerne

162 Two mourning women, 20 in.
DM Burch-Korrodi 1962

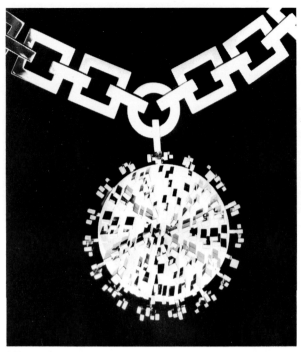

163

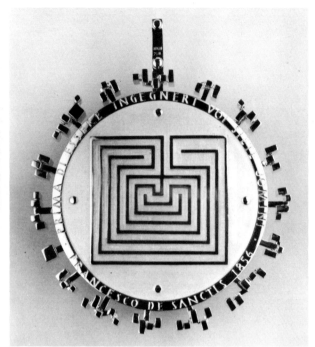

164

165

166

163/164 Gold medal for the Rector of
the Swiss Technical High School.
Inscription – You are men first,
engineers second. Symbolised
Knossos maze
DM Burch-Korrodi 1966

165 White enamelled sterling silver
chalice, inside gilt, symbol in
20 ct. gold
DM Burch-Korrodi 1964

166 Silver gilt sports trophy, the lion
of Zürich awarded by the City
DM Burch-Korrodi 1949

167 'Ultra' silver flatware
DM Sigurd Persson,
Stockholm 1953.

168 Stainless steel flatware for
Scandinavian Airlines System
(SAS), the winning design in a
Scandinavian competition. The
pieces are specially small to suit
the cramped eating space in
aircraft
D Sigurd Persson
M Kooperativa Förbundet 1959

169 Tea caddy, silver with blue
cloisonné enamel lid
DM Persson 1950

170 Tobacco box, silver with orange
cloisonné enamel on black
ground
DM Persson 1955

167

168

169

170

Three silver coffee pots showing how a small variation can change the character of a simple design:

171 **DM** Persson 1954

172 **DM** Persson 1960

173 **DM** Persson 1958

174 Silver thermos
 DM Persson 1962

175 **DM** Persson 1965

176 **DM** Persson 1963

177 **DM** Persson 1963

174

175

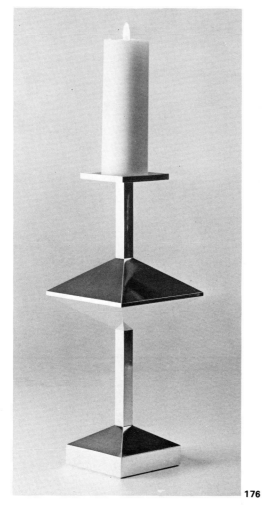

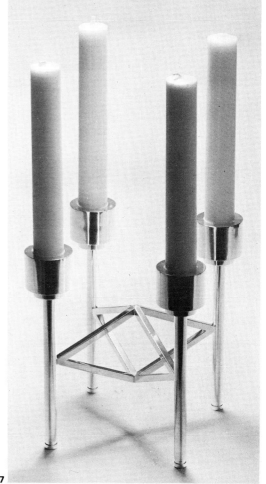

176

177

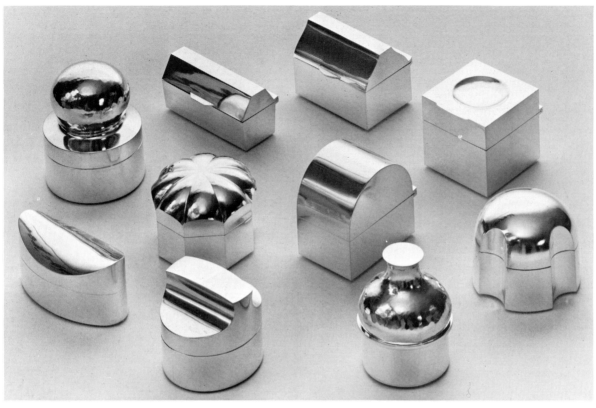

178 Silver boxes
  DM Persson 1966

179 Communion plate from the new
  Thomas Church, Vällingby
  DM Persson 1960

180 Chalice for Hedemora Church
  DM Persson 1960

181 Sigurd Persson's one man show
  'Silver Candlesticks' in the
  Hantverket showroom, Stockholm
  1964.

179

180

182 Altar set for Olaus Petri Church,
Gärdet, Stockholm: silver with
hammered gold strips
DM Persson 1959

183 Altar set for Adolf Fredriks Church,
Stockholm
DM Persson 1960

184 Crucifix, Birgitta Church,
Stockholm: gilt wood figure in
cast silver
DM Persson 1961

185 Processional cross, silver and red
glass, for Strängnäs Cathedral
DM Persson 1957

186 Silver and gilt, avanturine and
lapis lazuli, Hälsingborg
Crematorium
DM Persson 1961

187 Brass, pine wood and silver, for
Säffle Church, in Värmland
DM Persson 1964

185

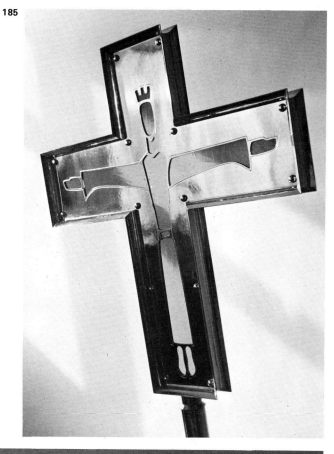

186

187

189

188

188 Silver cutlery with gold ends
  DM Gerald Benney, London
       1960
  O The Ionian Bank, London,
       who commissioned this,
       perhaps the first set of
       specially commissioned
       cutlery of modern design in
       Britain since 1945

189 Stainless steel cutlery 'Studio'
  pattern, the first in Britain to be
  entirely machine-made with high
  relief pattern stamped into this
  very hard metal. Production in
  1967 is about 1500 dozen pieces
  weekly
  D Benney
  M Viners of Sheffield 1965
  O Worshipful Company of
       Goldsmiths

190 Beaker in Britannia standard
  silver (95·8 per cent pure)
  DM Benney 1962
  O Worshipful Company of
       Goldsmiths. Ht 4½ in.

191 Sugar Dredger
  DM Benney
  O Reading Corporation,
       presented by Reading Estate
       Agents and Auctioneers
       Association

191

190

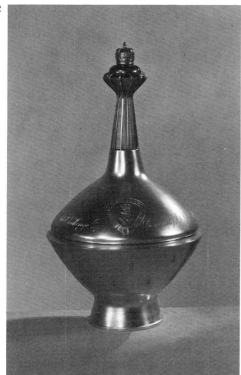

192

193

192 Silver Challenge Cup.
DM Benney 1955: Engraver
T. C. F. Wise
O Royal Perth Yacht Club,
Australia, presented by Royal
Thames Yacht Club, London

193 Detail of gilt and oxidised silver
crown in plate 192

194 Pepper mill, ht 6¼ in.
DM Benney
O Christ's College, Cambridge
Engraver T. C. F. Wise, 1963

195 Triangular gold cup with
palisander wood plinth and black
leather box. Awarded annually
with a miniature replica, for
outstanding motoring achievement.
Winners include Dunlop, Coventry
Climax Engines (makers of the
Lotus) and Jack Brabham. Like
many of Benney's pieces, this is a
good example of industrial
patronage of art. The donors,
Ferodo, publicize their own name
and products and earn goodwill at
the same time as encouraging
merit.
DM Benney 1959
O Ferodo Ltd

194

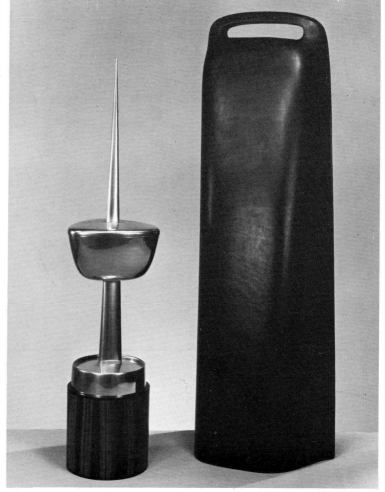

195

196

197

198

196 Gold cup, won at the Epsom
Derby on 6 June 1962. Such a
simple trophy for a world-famous
race is, alas, extremely rare:
sports trophies are almost always
flamboyant and vulgar
DM Benney

197 Silver trophy commissioned by
the Duke of Edinburgh for the
annual athletics meeting between
Edinburgh and Munich. The name
of one city is on each end and
the winner lives on top. The
chased herring-bone pattern
invented by Benney brings
richness and colour to sporting
trophies where austere modernity
is usually unwelcome.
DM Benney 1966

198 **DM** Benney for his own use
1952

199 Large teapot, triangular section
Length 11¼ in.
DM Benney 1959
O Graham Hughes

199

200 Pair of gilt water jugs
commissioned by local
subscription and given to
Stamford Corporation to celebrate
its 250th anniversary, 1957
DM Benney. Ht 12 in.

201 Water jug. Designed and made
in five days by Gerald Benney
for the Worshipful Company of
Goldsmiths' film *A Place for
Gold*, 1957. A more normal time
for the creation of such a piece
might be a month. Ht 10 in.

200   **201**

117

202 Commissioned by St John's
College, Cambridge, to celebrate
its Master Revd J. S. Boys Smith
becoming Vice Chancellor of the
University
DM Benney 1965. Ht 9½ in.

203 Silver with gilt feathers applied
DM Benney 1963. Length 7 in.
O Worshipful Company of
Goldsmiths

204 Textured sauceboat
DM Benney 1966

205 Pair of large water jugs, part of a
gift by the Worshipful Company
of Goldsmiths to the new
University of Sussex at Brighton
DM Benney 1965

204

205

206

207

208

206 Small gilt boxes, from a long series showing Benney's perpetual delight in texture and ornament, and his mechanical resourcefulness in using simple tools to achieve precise effects
DM Benney 1963. Length 4 in.
O Worshipful Company of Goldsmiths

207 One of Benney's first big opportunities, already showing exuberant invention coupled with a strong underlying shape. Box of silver, ornamented with appliqué ovals, each with a ball of gold in the centre. The lid of mahogany, inlaid with strips of silver. The coat of arms chased and engraved in gold and enamel. Inscription inside lid refers to the ancient Bristol guild: 'Presented to the Society of Merchant Venturers by Foster Gotch Robinson, Master 1943 and 1944. Designed and made by Gerald Benney, 1955' Engraver T. C. F. Wise: enameller R. E. Tyler. Length 13 in.

208 Benney's new ideas regularly replace his old. These gold and silver boxes are his latest range. They are extremely difficult to make because the parallel strips of metal react differently to heat and therefore the surfaces tend to warp under manufacture. The pattern is a new use for, and may lead to a revival of, the old technique of chasing.
DM Benney 1965
O Worshipful Company of Goldsmiths

209 Scroll holder, a new type of container invented by Benney for an old type of paper. Commissioned by the Institution of Electrical Engineers for presentation to HRH The Duke of Edinburgh on his admission as an Honorary Member, 1956. Engraver: T. C. F. Wise. (The engraving symbolizes electricity)

210 Scroll holder in silver and wood, commissioned by the Borough of Aldershot for the Parachute Regiment on its receiving the Freedom of the Borough in 1957
DM Benney: engraver T. C. F. Wise

211 Freedom casket commissioned by the Borough of Aldershot for the Corps of Royal Engineers on their receiving the Freedom of the Borough 1965
DM Benney, engraver T. C. F. Wise

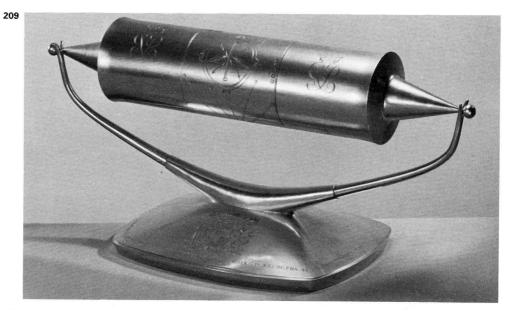

209

210

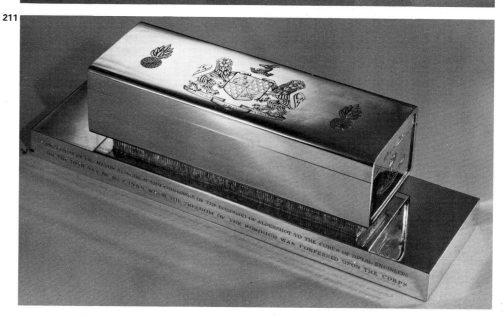

211

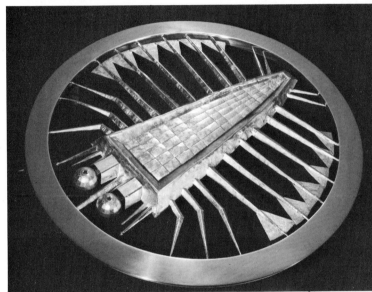

212
214
216

213
215
217

122

212 Large silver bowl with pierced
cover and gilt inside. Given to
the new Leicester University by
its first Chancellor Lord
Adrian 1958.
The beginning of a fine series of
table centrepieces by Benney,
sometimes used to hold flowers,
sometimes, without the cover,
for fruit or bread. A few such
bowls were made in Britain
before 1939 by silversmiths like
Murphy and Omar Ramsden
who called them rosebowls,
referring to the big dishes of the
16th and 17th centuries in which
rose petals floated, and in whose
perfumed water one washed
one's fingers
DM Benney
(See also back cover)

213 Inscribed beneath the base:
'Presented by the Directors of the
National Bank of Scotland
Limited to the Rt. Hon. Lord
Rowallan, KT., KBE., MC., TD.,
LI D., their Governor, in recognition
of his outstanding services to
the Bank from 1951 to 1959.'
The inside of the bowl, and the
upright rings in the centre of the
cover are gilt
DM Benney 1959: engraver
T. C. F. Wise

214 Silver parcel-gilt, with rhodium-
plated central sunflower
presented by Courtaulds Ltd to
Professor D. M. Newitt on his
retirement from the Courtauld
Chair of Chemical Engineering at
the Imperial College of Science
and Technology 1961
DM Benney. Diam. 12¾ in.

215 'Beetle' cover, inspired by an
underwater microphotograph of
bacteria, 22 ct. gold eyes set
with peridots. Diam. 20 in.
DM Benney 1962
O Worshipful Company of
Goldsmiths

216 Bowl, silver and gold. Given to
the Royal College of Physicians
by Professor R. A. McCance
DM Benney 1962

217 Scorpion cover
DM Benney for Meister
retailers of Zürich 1964

218 Pair of bowls with covers.
Given by Mr Charles Clee to
London University 1959
DM Benney. Diam. 14 in.

219 Presented by Sir John Cockroft
to Churchill College, Cambridge
DM Benney 1960

218

219

220

221

222

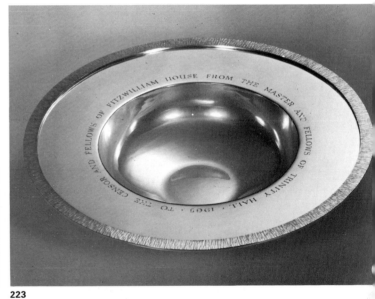

223

220 Given to Ipswich Corporation
Suffolk by the Goodchild family,
whose father Edwin Goodchild
was born there, lived there for
65 years, and emigrated to
South Africa
DM Benney 1963

221 Large bowl given by the world's
largest bullion company,
Johnson Matthey & Co. Ltd of
Hatton Garden, London, to their
US associates, Universal Oil
Company, Chicago, to celebrate
their 50th anniversary 1914–1964
DM Benney 1964

222 Table jewel, silver parcel gilt
with topaz in centre
DM Benney 1964

223 Given by Trinity Hall, Cambridge,
to Fitzwilliam House to celebrate
its becoming Fitzwilliam College
1965
DM Benney

224 Large silver bowl with pierced cover, stones and gilt inside. Given by Standard Industrial Group Ltd to André Felix de Breyne
DM Benney 1967

225 Chalice, gilt inside, with hole in knop. Probably the world's first piece of silver since the 18th century to be made with an original textured surface of a type now becoming popular with several silversmiths not only for beauty but also to prevent finger marks. Benney applied this particular texture by engraving, but subsequently used chasing or hammering. Commissioned by the Worshipful Company of Goldsmiths, and sent by them to tour USA with the important exhibition 'British art in craft'
DM Benney 1956

226 Ciborium with finial referring to the cross of nails which survived the German bombing
DM Benney 1958
O Coventry Cathedral

224

225

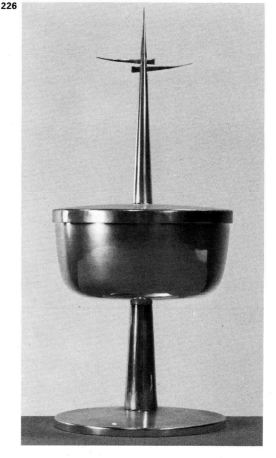

226

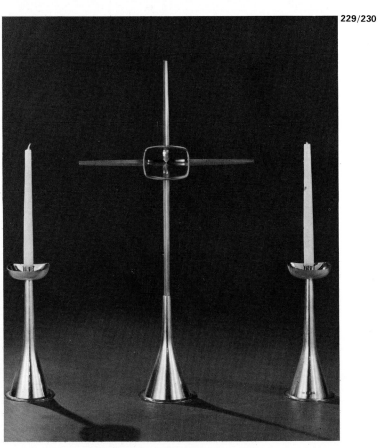

227/228 Altar cross, parcel-gilt, with rough and smooth surfaces to catch the light and stand out from the Victorian background masonry
DM Benney 1963
O Charterhouse School Memorial Chapel, Godalming, Surrey

229 A shape of our times to give modern appeal to a gloomy Victorian apse. The central tongue is gilt.
DM Benney 1960
O Gonville and Caius College, Cambridge

230 Part of a fine group of modern church work by various artists
DM Benney 1958
O Bishop Otter College, Chichester, Sussex

231

231 Altar cross with rays and candlesticks, given by the Normanby and Phipps families 1964
DM Benney
O Lythe church, Whitby, Yorks

232 Modernity for one of the finest medieval churches near London
DM Benney 1964
O Harefield church, Middlesex

233 All Benney's altar plate is designed to suggest light and supernatural movement and it is often textured, to contrast with its background and achieve powerful impact on worshippers. One of the extremely few pieces of Jewish ritual plate to deviate from the traditional style. Eternal light for New London Synagogue
DM Benney 1966

232

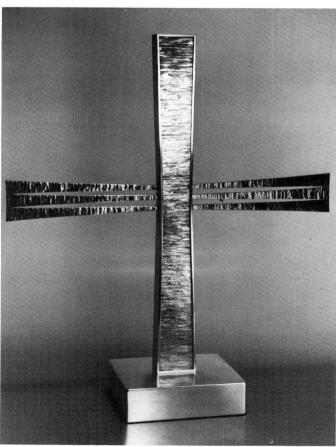

233

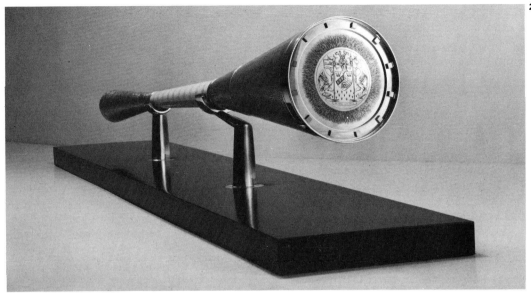

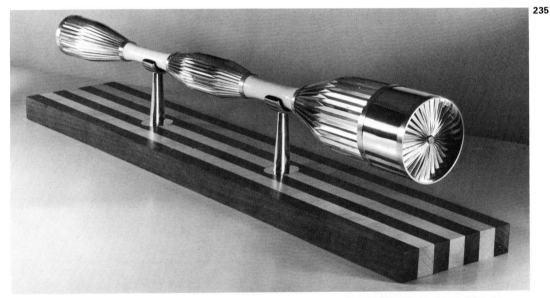

236

234 Mace in silver parcel-gilt and
ivory, given by East Sussex
County Council to the new
University of Sussex at Brighton
1963

235 Mace in silver and ivory, given
by Norwich Junior Chamber of
Commerce to the new University
of East Anglia at Norwich.
Christopher Lawrence, one of
Benney's assistants, won the
Cartier Memorial Award at
Goldsmiths' Hall 1963 for his
particularly intricate and difficult
silversmithing on the creased
surfaces. Length 39 in.
DM Benney 1965

236 Mace
DM Benney 1965
O University of New South
Wales, Newcastle, Australia

Patrons and design

Church, companies, societies, universities,
competitions. The evolution of style 1890–1914,
1918–39, 1945–67

*It is not this or that tangible steel and brass machine
which we want to get rid of, but the great intangible
machine of commercial tyranny, which oppresses the lives
of all of us. Now, this enterprise of rebelling against
commercialism I hold to be a thoroughly worthy one: our
aim should be to add to the incentive of necessity for
working, the incentive of pleasure and interest in the work
itself. I am not pleading for the production of a little more
beauty in the world, much as I love it, and much as I
would sacrifice for its sake; it is the lives of human beings
that I am pleading for.*

*Applied art is the ornamental quality which men choose
to add to articles of utility. Theoretically this ornament can
be done without, and art would then cease to be 'applied'
. . . would exist as a kind of abstraction, I suppose.*

*But if these applied arts are necessary, as I believe they
are, to prevent mankind from being a mere ugly and
degraded blotch on the surface of the earth, which without
him would certainly be beautiful, their other function of
giving pleasure to labour is at least as necessary, and, if
the two functions can be separated, even more beneficent
and indispensable.*

William Morris, addressing the National Association for the Advancement of Art,
in Liverpool and Edinburgh in 1888 and 1889

Design is spontaneous in those objects which are still made by hand, such as painting, sculpture, or women's hats. The moving force is fashion, and it is amazing to contemplate its power: the Rococo from France swept Europe in the 1730s almost as quickly as the mini-skirt from England today. The need for change seems fundamental to human life.

Fashion was and still is the accelerator: one of the strongest human instincts is to keep up with people one admires, and usually these are more wealthy, more enterprising, more beautiful, more susceptible to change than oneself. Machinery became the brake. Machinery stopped the swing of the fashion pendulum in the nineteenth century and, while any machine-made object of truly mass demand, such as a motor-car, may change year by year to titillate the tired driver's palate, more specialized luxuries, like silver, for which there is less demand, will remain the same. The financial incentive to produce a new pattern of cutlery, which may cost the factory an investment of £15,000, is insufficient because of the small size of the luxury market. So while human ingenuity has devised ways of continually stimulating sales by changing many other hand and machine-made objects, for which the demand is almost universal, machine-made silver remains constant because so few people want it. In 1967 smart shops the world over are offering for sale designs that were already common in 1930, if not in 1867.

So it is hand-made pieces in their dozens, not factory work in its millions, that show the sensitive artist responding to his own antennae; and that is why many art historians concentrate so much on relatively restricted movements like *art nouveau*, which hardly affected the main channels of commerce.

In retrospect, we may see that the back of London's St Pancras Station, the magnificent and original train shed of 1867, which could have been built at no other time and in no other country and which represented a vital new invention – mass communication by railway – matters more than the front, a 're-do' of Gothic which was better done 700 years before. At the time, it didn't seem so simple: the back was just functional. The front was fashionable and called 'Art'. The struggle between function and fashion was not then recognized: it came to fruition in the twentieth century, not the nineteenth, and nowhere more vividly than in silver.

We like to think our judgement is improving, and aesthetes often talk of good and bad design. What they mean is, 'I like it' or 'I don't'. James Laver, the wisest and wittiest historian of fashion, in his splendid book *Victoriana*, shows that, although some designers will achieve magnificent technical improvements, fashion, not some high ideal, is usually the dominant motive in applied art.

Forty years ago, the word 'Victorian' was simply a term of abuse; it stood for all that was stuffy, heavy, and overladen with ornament. Lytton Strachey, in his wickedly urbane Eminent Victorians *had just demonstrated that the Victorian Age was not only all these things but ridiculous as well . . .*

This raises one of the most difficult questions in the psychology of taste. We like to think that taste is static, and what we admire today would always have been admired if only our ancestors had been educated up to it. We like to think that beauty and ugliness are objective realities, eternal in the heavens, unchanging although the skies should fall. The truth is, unfortunately, very different.

It is very instructive in this connection to study the history of feminine fashions; that is, the department of the applied arts which is most easily and most visibly, influenced by 'Fashion'. It seems to be a law of our own minds that the fashions of our mothers are hideous, the fashions of our grandmothers quaint, the fashions of our great-grandmothers charming, and the fashions of our great-great-grandmothers beautiful. Anyone can check this for himself by looking through a historical sequence of fashion-plates. But the same law operates, at a slightly larger remove, in all the other applied arts. It is equally true of furniture and interior decoration. There is an incapacity to appreciate what lies immediately behind us and a readiness to accept (and even to pay highly for) what lies further back. The 'second-hand' and the 'old-fashioned' gradually take on the patina of the 'antique'.

It almost seems as if there were a Gap in Appreciation *stretching across a given number of years, and that the inevitable way for furniture dealers to make a fortune would be to buy up everything they could lay their hands on a few years ahead of public taste. Indeed, this is precisely what has happened. Early collectors of Chippendale and Empire furniture have made fortunes, just as early collectors of Victoriana knick-knacks have already begun to do. One ought to lay down furniture for one's children just as our ancestors laid down wine, in the firm conviction that with the passage of the years it will inevitably mature. . . . The Gap in Appreciation not only enables designers to throw off the burden of the immediate past and to move forward to the creation of a new style, it* spaces out *the surviving objects, and so makes it possible to collect them.*

1890–1914

The period about 1890 begins with a conscious revival of original design; patrons were tired of poor quality factory reproductions from antiquity and William Morris was more than tired: he was angry. He wanted workers to enjoy their 'daily necessary work', and he wanted artists to express their own inner nature, not copy someone else's. In 1891 he started the Kelmscott Press to launch his fine printing, a specialized offshoot of his own firm Morris & Company, 'the firm'. This lasted till 1940, producing his own superb and creative fabrics and wallpaper but, alas, otherwise selling uncreative and repetitive metalwork, mostly concentrating on the imitative talent of the Pre-Raphaelite Brotherhood. This had been founded in 1848 by Holman Hunt and John Everett Millais, one of the most gifted of all British artists; and Millais, in 1854, had married Effie Gray, formerly John Ruskin's wife. So Morris commanded what he thought was the best talent of the age and harnessed it to the best theory, namely his own and Ruskin's. He set in motion much more than he knew: the whole modern movement; his message succeeded in the wide world although his own firm failed to transform England.

He rightly felt that humble workmen matter, but he wrongly tried to give workmen back their self-respect through using their hands, instead of concentrating on the machine, which was with us to stay whether Morris liked it or not.

Before 1914 the buyers who stimulated design were still the private owners, often the great aristocrats who still maintained their country houses in feudal splendour. The young married couple with such a background would usually be equipped with table settings enough for fifty people even though when they inherited the title of duke or earl they knew that they would also inherit the family table settings for two or three hundred. England was still the workshop of the world and orders from exotic foreign potentates came to the big Sheffield and Birmingham factories almost as a matter of course. Support from the nobility at home and abroad was then crucial to prosperity.

Most commissioned work, however, was probably not for private homes. This was the last epoch during which the visionary dream of unity between religion and art still applied. Church buildings arose in great numbers in every country, to keep up with the increased population, and many different shades of religious feeling required different types of ritual and ornament. Some priests, for instance, liked the figure of Christ on the altar cross, some liked three candles each side of the cross instead of one, sometimes incense burners and processional crosses were prominent. Church money often came from public subscription, and there was so much of it that, for the first time, firms of professional church furnishers arose. This was the heyday of Barkentin and Krall 1861, the Warham Guild 1912, associated with the Church leaders and others, Wippels, registered at Goldsmiths' Hall 1903, Blunt and Wray, registered by Sidney Blunt and Frederick Wray 1888, Burns Oates and Washbourne, registered as R. and T. Washbourne 1906, Mowbray's 1924, and Faith Craft Works 1938. The firms often had the best intentions and nearly always had accurate, admirable and specialized knowledge of the Christian background. But they nevertheless stood between the artist and the act of worship. Professional church furnishers were only middle men. Because their products were easy and cheap to buy, and because they worked well, they drove out the original creations of the late Victorian architects like Sedding, Butterfield and Henry Wilson. This trade in religious art was too successful. Real art shied off, and in the next generation was seen more in houses and in galleries than in churches; religion too was damaged because its surroundings became stereotyped.

The most delightful metalwork of the time was either jewelry in the *art nouveau* style, or wrought iron architectural ornaments such as those of Victor Horta which may be seen today in Brussels at the Hotel Solvay, the delight of whose doorknobs keeps the visitor continually on the move; those in the Glasgow School of Art by Charles Rennie Mackintosh which started a new and alas only temporary taste for modern iron cusps and finials, the Barcelona houses by Gaudí with intricate swelling balconies, or the Paris métro entrances by Guimard like tall black metal flower gardens. Beside these original masterpieces, church and table silver alike seemed uneasy.

On the one hand is the hangover from the hard Victorian Gothic style – elaborately cast and saw-pierced, much admired for its intricacy at the great London exhibitions of 1851 and 1862, but by 1900 recognized as artistically dead. On the other hand were the reform movements: the Arts and Crafts in England which stimulated new life everywhere, *art nouveau* in France and Belgium, Jugendstil in Germany, Stile Liberty or Floreale in Italy, Sezession in Austria. All of these produced exceptional ideas, and some of them had a big influence on the development of design.

Jewelry and graphic design, fabrics and architectural metalwork were the forms in which *art nouveau* swept the world. In the fine arts the effect was almost imperceptible. Klimt is the famous *art nouveau* painter, but the style never affected his much greater contemporaries like Van Gogh or Cézanne, and is never even mentioned in the memoirs of Renoir by his son Jean, otherwise such a comprehensive record of the times. One can walk through the numerous cities of Europe being enlarged at the height of the craze, even the centres like Munich, Vienna, Paris or Brussels, without seeing a single building in the style. With silver, too, it never covered more than a tiny fraction of the whole productive effort. Most factories stuck to their eclectic inheritance, and either went modern when they invested in new machines after 1918 or simply dwindled away and died.

Successful modern silver producers were, in fact, so few that they could easily be listed: Jensen of Copenhagen, Kaiserzinn, who made pewter in Cologne, and Liberty of London were some of the big names who harnessed sentiment to reality, and each of these producers developed his own distinctive identity. They did this so emphatically that they later found it difficult to change. Only Jensens are still an active force in silver today, and even they have a struggle to outlive their past.

For some historians today *art nouveau* represents an irresponsible shrug of the aesthetic shoulder, for others a wild and degrading craze, for others a serious universal movement like Gothic or Baroque, which lasted so short a time, not because it was trivial but because of the speed of modern communications. It is too soon to make a final judgement, and it does not really matter. What is certain is that the imaginative power of these artists was astonishing: in applied art their products have fascinating variety, spontaneity and romantic abandon.

Only some countries were affected, and only some designers in these countries. The impact was terrific in its intensity, but limited in its scope. *Art nouveau* was suited to the boudoir, not to the factory bench, to handwork for wealthy individual clients, not to mass-production for utility. Many prototypes were made in fabric, wallpaper, glass or silver, but with the exception of the British firm, Liberty, the intended quantity production did not usually take place: industry maintained its steady progress towards efficiency of form and function, and declined the exotic fantasy which some of the leading designers' models offered them. It was the old crafts, where most of the best work was still done by hand, like silver and jewelry, that could and did adapt themselves most radically, and it was the art which has the fewest rules of all, namely jewelry, that gave the style its most splendid monument.

Each country had its own particular style, often with a national name. The whole movement was so short-lived and created by such a small number of artists, each of whom remained a leading influence in his own group, that national ideas remained undiluted to the end. But there were international exchanges on an inspiring scale which could only have been achieved by real idealism and romantic imagination. The international exhibitions in Philadelphia 1876, New Orleans 1884–5, Paris 1867, 1889, and 1900, Chicago 1893, Brussels 1897, Munich 1897, Dresden 1897, Turin 1902, and St Louis 1904 showed the world's instinct for unity and prosperity. Ashbee of England exhibited at Vienna, Munich, Düsseldorf, Paris, and designed rooms for the Prince of Hesse's castle at Darmstadt, destroyed by bomb damage in the 1939 war. Mackintosh of Glasgow designed the brilliant Scottish Pavilion in the 1902 Turin Exhibition, and showed also in Vienna, Budapest, Dresden, Munich and Moscow. Joseph Hoffmann of Vienna designed and built his masterpiece not there but in Brussels: the Palais Stoclet. Van de Velde was born in Antwerp, died in Zürich, worked in Germany, Switzerland, Holland and Belgium. Tiffany studied in Paris and won a Grand Prix at Turin 1902. The St Louis 1904 exhibition represented many of Europe's leading names. Wolfers of Brussels had jewels reproduced in eighty-three art magazines all over the world 1893–1908, and the showcase he designed, once at the Paris 1902 exhibition, is now appropriately in the Darmstadt museum. The different artists showed an intense interest in each other's work without ever apparently wanting to copy it.

For Britain it may be said here, as in other fields, that she invented it but never quite got there. A. H. Mackmurdo (1851–1942) started the Century Guild in 1882 to drive trade out of the arts, and his 1883 book cover is the world's first recorded *art nouveau* design, but he abandoned the lead he had won.

C. R. Ashbee (1863–1942), another architect and the most prolific metalworker of the time, started the Guild of Handicraft in 1888, as a craft co-operative, at Essex House in the Mile End Road, moved it in 1902 to Chipping Campden in Gloucestershire where old houses were restored and agricultural work undertaken, retaining a London shop, and finally wound it up in 1908. Ashbee seems now to have been one of the quieter, less creative designers, but his reputation then was good: in 1912 his *Silverwork and Jewellery* privately printed, showed his interest in, but also his rather limited appreciation of craft work. His jewels were usually clumsy and unfeminine, made of silver and one of the translucent stones, blister pearl, moonstone, turquoise or opal, but always much more metal than stone; and his silver followed suit.

Next in importance probably comes the firm of Liberty who were started in 1875 by Arthur Lasenby Liberty partly to popularize unsold Japanese stock remaining from the first western showing of Japanese work, the London 1862 exhibition. William Morris managed the Farmer & Rogers oriental warehouse and inspired his friend Liberty with zeal for Japan, whose influence on *art nouveau* was partly due to the novelty of the country, opened to the West in 1857. Liberty in his turn used many metalwork designers, his chief suppliers probably being William Hutton & Sons of Sheffield and W. H. Haseler & Co. of Birmingham; in 1899 he launched the 'Cymric' range of silver and jewels, still commonly seen in British antique shops, stamped with the trade name, heavily worked, soberly but not brilliantly eccentric. 'Tudric' pewter followed. Unfortunately the Liberty policy was to keep the names of their designers secret, no doubt in order to retain control over them; this means that, despite splendid researches by Mrs Shirley Bury at the Victoria and Albert Museum, we don't know what proportion of the work was machine-made, or how far the firm either wanted or achieved a union of art and industry. Archibald Knox (1864–1933) was one of the more original designers, Bernard Cuzner (1877–1956) a typically devoted craftsman who looked back on his *art nouveau* work as a youthful irresponsibility, a whimsical betrayal of the solid materials he admired. Haseler himself, conversely, probably valued production for its own sake: his joint company with Libertys was dissolved as late as 1927.

Charles Rennie Mackintosh (1868–1928) is now recognized as the world giant of the period, the one British designer whose invention never flagged, whose conviction was absolute, whose stature is comparable with Van de Velde and Lalique. He designed tableware prototypes and architectural metalwork which have excited stylists everywhere, and those numerous austere art theorists who, as valid twentieth-century art forms, prefer doorknobs to jewels. Mackintosh was the catalyst for a group of artists who came to be centred on the Glasgow School of Art, including his wife Margaret MacDonald (1865–1933), and her sister Frances (1874–1921), Jessie King (1876–1949), Nelson Dawson (1859–1942), Talwyn Morris (1865–1911) and others, all of whose jewels and metalwork are much commoner and less distinguished than their master's, though still dominated by his own personal idiom, as astonishing as it is logical.

If Ashbee made the most jewels, Henry Wilson (1864–1934) led in silver. Professor Pevsner surprisingly claims that Ashbee's silver is not *art nouveau*, in which case Wilson's, with its castellated romantic medievalism, is even less so. Both rather homespun stylistically, Ashbee was more designer, Wilson more craftsman, though he himself might not have liked the assessment, as he was an active author and lecturer. He made great quantities of church plate and regalia, for health reasons often doing his casting at Torcello. He made the enormous doors for New York Anglican Cathedral. He is a link with both the past and the future, because his style is almost so backward-looking as to be Victorian, almost progressive enough to confuse with Omar Ramsden and Alwyn Carr, the next generation of church smiths. Wilson was a sensible serviceable bridge between the nineteenth and twentieth centuries, with Voysey's or Dresser's austere unornamented functionalism at one extreme, and the exuberant *art nouveau* at the other. What he carried over the bridge is not now fashionable: it is William Morris's belief in the joy of handwork. Yet Morris's energies were not in vain because England is now the only country in the world where hand-made silver is still often commissioned from independent artist-craftsmen.

British *art nouveau* was almost stillborn because of British reticence; as a nation we do not indulge in orgies of visual fun. Walter Crane (1845–1915) called the phenomenon a 'strange decorative disease' and 'the antithesis of the Morris school of decoration'. Morris himself (1834–96) was primarily a social reformer. Although his wallpapers and fabrics were commercially successful, his main artistic conviction was a sentimental one in favour of the individual as against the machine. He never reconciled the contradiction between his socialist and his artistic theories, the inability of expensive handwork to satisfy cheap mass markets to lighten the world's burden of poverty. The nearest he got was to praise sensible designs at the expense of fantasies. As he was a very powerful and popular personality his ideas, which dominated the applied art world, may have helped to undo British *art nouveau*, at the same time giving a world impetus to conscientious workmanship.

In Denmark there was the work without the theory. Georg Jensen and Johan Rohde quickly grew out of the prevalent style, imitative and over elaborate, and evolved their own personal idiom which has served the Jensen name so well ever since: more bulges than curves, more bulk than line, more fleshy than sinuous, more heavy fruits than light flowers, the idea was a steady success from the beginning. The material, as in Britain, was almost always silver, often with amber, the common Polish stone; and it may be guessed that jewels formed a much larger proportion of Jensen's output than they did of that of his contemporaries in England. Jensen designed with greater conviction than his British colleagues (and without the hampering influence of the Ruskin/Morris theories); moreover, precious jewelry and good silver hardly existed in Denmark before. In Britain, on the other hand, there were already hundreds of factories and retail shops: here the craftsmen's task was to convert an existing taste – a task which is only now nearing success.

In Norway too there was almost virgin ground. In the eighteenth century Bergen had been a great centre for filigree work, and Bergen smiths, such as the Reimer family, had exported all over the world. But it was not until the late nineteenth century that Norwegian silver, mostly from country workshops, was revived. The old firm of Jacob Tostrup pioneered not only filigree but a very fine, specialized transparent enamel – *plique à jour* – which won many international exhibition successes. The architect Torolf Prytz and later Jacob Prytz, head of the firm after 1918 and then head of the state art school, worked with the Oslo museum to bring a new vitality to applied art. Other firms like David Andersen followed, until Norway became widely famous for its modern enamels.

In Germany 'Jugendstil' (from the Munich paper) was also a commercial success. In fact it was commercial from the start, and therefore different in detail, being adapted to semi-quantity production. Machine-made components were assembled in small, already existing factories, mostly at Pforzheim, since the eighteenth century the great centre for gold and silver jewels, but also at Frankfurt, Schwäbisch Gmünd and at Hanau. The products were small, unlike French and Belgian work, light, unlike British or Danish, and anonymous, unlike any other country's. The great German pioneers of international modern architecture, Peter Behrens, Walter Gropius and Joseph Olbrich killed the curve almost before it had appeared there. But, as in Denmark, German work around 1900 led to enormous commercial expansion in the 1920s and 1930s: the new Pforzheim Reuchlinhaus Museum has the most elegant jewelry display in existence, and shows the development from the sophisticated pioneer, small scale production of 1900, to the vulgar meretricious work which today has gained for Pforzheim its undisputed place as the world's leading producer of cheap work.

Individual German metalwork designers like Karl Bauer (1868–1942) or Georg Kleeman (1863–1929) or Ernst Riegel (1871–1946) did not apparently have very personal styles, and the giants like Behrens or Riemerschmid, and indeed the whole artists' colony established at Darmstadt, had more influence than the craftsmen themselves on silver design. The Darmstadt museum, brilliantly rebuilt and completed in 1965, shows this German Jugendstil, in many media, at its best.

Vienna was one of the first capitals to meet *art nouveau*, but one of the last to digest it. Of the founders of the anti-historical 'Sezession' group in 1897, which gave its name to Viennese Jugendstil, Joseph Hoffmann (1870–1956) and Koloman Moser (1868–1916) both designed silver. Many of the leading artists, including Hoffmann and Dagobert Peche (1887–1923), supported the Wiener Werkstätte, started by Hoffmann in 1903, a society of artist-craftsmen with furnished workshops which survived till the 1930s slump, having made, over three decades, much of the best Viennese craftwork. The style was that of 'chequerboard' Hoffmann, small squares and triangles and dots, much richer than the German, often using gold or pearls or ivory, or enamels, often producing rather large pieces for the wealthy classes. The great achievement is Hoffmann's Palais Stoclet in Brussels where many of the craftsmen worked 1904–11, and where some of the best of the rare Sezession pieces are still preserved.

It was in Belgium that silver in the 'modern style' as it was called, first became fashionable, through the influence of Philippe Wolfers (1858–1929), the very substantial artist whose firm Wolfers Frères were and still are the Belgian Crown Jewelers with all the power which that responsibility carries. It must have been this power, so seldom well exercised by its holders, that enabled Wolfers to produce a long series of very precious jewels, the finest group of which is now owned and venerated by the firm and family. He studied at the Beaux Arts in Brussels, joined the family firm which, however, had no retail shop, was much impressed with Japanese work at the Vienna International Exhibition 1873, *c.* 1890 established his own workshop in Marie Louise Square, *c.* 1893 started using ivory from the Congo offered to artists by Leopold II. From 1905 he became increasingly interested in sculpture, abandoning jewels, his first love, and interior decoration, his second, and in 1910 Victor Horta built the firm's splendid workshops and showrooms.

Henri van de Velde (1863–1957) was born in Antwerp, worked as architectural adviser to the Grand Duke of Weimar 1899–1917, then taught at Ghent and Courtrai, also working intermittently in Holland and in Switzerland. Unlike Mackintosh, with whom his strength and originality are comparable, he must have particularly enjoyed silver and jewels: his surviving pieces are almost all of austere silver or base metal – architect-jewelers don't seem to respond to the joys of luxury – but Van de Velde's work makes up in exquisitely elaborate linear chasing what it loses in intrinsic lustre. He strikes the perfect balance between an architect's perfectionism and a jeweler's fantasy, a rare combination of qualities which gives his metalwork unique distinction.

The name *La Maison de l'art nouveau* was invented by S. Bing of Hamburg for the shop he opened in Paris in 1895, from which he must have sold much of the silver under discussion. Always a jewel centre, Paris was seething with change at the turn of the century, much more so than London, its sister in wealth and politics. Impressionism in art, Proust, Victor Hugo, De Musset and symbolism in poetry, Diaghilev, Stravinsky, Debussy and Ravel in music, the Franco-Prussian War of 1870 and the Haussmann replanning on a huge scale afterwards, made women ready for a change, and indeed they expected it. René Lalique (1860–1945) is the most sensational figure in the field though he designed almost no silver. He always wanted to draw. Son of a merchant, on his father's death in 1876 his mother apprenticed him as jeweler to Louis Aucoc, and he started courses at the École des Arts Décoratifs, which he abandoned for lack of time. From 1878 he studied at Sydenham, and in 1881 returned to work with various Paris firms, designing wallpapers, fabrics, and an industrial art journal with etchings as a guide to jewelers. By now he knew some of the customers of the firms for whom he had worked – Cartier, Boucheron, Aucoc, Destape – and in 1886 he was left Destape's business. From 1890–3 he studied glass techniques in his Rue Thérèse workshop, and, equally important, attracted Sarah Bernhardt who commissioned two groups of jewels for *Iseyl et Gismonda*. In 1895 his first display at the Salon made him well known – he won third prize there. The same year he opened his Place Vendôme shop and first used the female nude in his jewelry. In 1896 he first used horn encrusted with silver, winning second prize. In 1897 he became Chevalier of the Légion d'Honneur. In 1903 he designed and built his shop at 40 Cours Albert I, with a great doorway in glass and pinewood, by Saint Gobin. In 1909 he leased and in 1910 bought a glass factory at Combes-la-Ville, glass becoming more prominent in his jewels, till he finally abandoned jewelry in 1914. Eugène Feuillâtre (1870–1916) worked for him, researching into the problems of enamelling on silver, and Lalique himself was a great technical innovator, using machines to reduce his large models to actual working size.

Georges Fouquet (1862–1957), the eldest son of Alphonse Fouquet jeweler of Avenue de l'Opéra, had a stylistic affinity with Lalique, and seems to have operated on a similar scale. We know of him from his family because his son Jean is now one of the foremost Paris artist-craftsmen. Georges started work with his father at eighteen; in 1895 he inherited the business and immediately modernized it, working with Tourrette, Desroziers (1905–8), Grasset and Mucha. After prizes in the Universal Exhibitions of 1900 and 1901, he asked Mucha to design his new shop at 6 Rue Royale, which was demolished 1920. He showed in Milan 1906, was president of the jewelers at the Paris 1925 and 1937 exhibitions, and a member of the Union Centrale des Arts Décoratifs, thus being one of the few designers to remain in practice having outlived the epoch. His most significant connection was no doubt with Alphons Mucha, the most prolific designer of the age, if not the most original, to whose designs he made many pieces, and through whom he must have met some of the most exotic women in the world.

Henri Vever (1854–1942), author of the history of French nineteenth-century jewelry, inherited his firm in 1874 with his brother Paul (1851–1915) from their father Ernest, a figure of standing who had been President of the Chambre Syndicale. Henri studied at the École des Arts Décoratifs in the evenings, and by day at the ateliers of Loguet, Hallet and Dufong. In 1889, at their first exhibition, the brothers won one of the two Grands Prix; in 1891 at the French exhibition in Moscow, they won the Croix de la Légion d'Honneur, and studied the crown jewels of the Tsars there, and of the sultans in Istanbul. In 1893, Henri was Commissioner at the Chicago exhibition; in 1895 he won prizes at Bordeaux, in 1897 at Brussels, in 1900 a Grand Prix at Paris. Vever's work was much heavier and less spontaneous than Lalique's or Fouquet's and he probably did not pride himself as the others did on producing outstanding special masterpieces.

So much for the biggest French names: foremost among the others was Alphons Mucha (1860–1939), painter, graphic designer and decorator, the best known and most prolific of all. He started his studio with Whistler. From 1894 he had a six-year contract for Sarah Bernhardt's posters and decors and published portfolios of designs which were disseminated so far that 'le style Mucha' and *art nouveau* were for a time synonymous. He worked in Paris, Berlin, and his native Prague, and designed rather lightweight jewels for Georges Fouquet from 1898 to 1905. He was more facile than inspired. Then there was Lucien Gaillard (b. 1861), known for his interest in Japan, who exhibited at the Paris 1900 and the Glasgow 1901 exhibitions; Eugène Grasset (1841–1917), another Japanese enthusiast, who visited Egypt in 1869, studied with Viollet-le-Duc and designed for Vever, more popularizer than creator. Victor Prouvé (1858–1943) pioneer of exotic glass made at Nancy under the inventive lead of Emile Gallé (1846–1904), Paul Liénard (1849–1900), Georges de Ribaucourt (1881–1907), L. Gautrait, Henri Dubret and others hardly established their personalities in this most personal of styles. All these artists followed the giants, but *art nouveau* was a medium for creators, not for followers.

Smart shops for this metal and glass existed in Paris as nowhere else in Europe. Lalique's was in the Place Vendôme where the group La Haute Joaillerie de France now mostly are, Fouquet's, designed by Mucha, was in the Rue Royale near the present shop bearing Lalique's name but not his philosophy. Some of the Fouquet-Mucha parts preserved in the Musée Carnavalet (but not on show) give an idea of the luxuriant fittings which must have made the exhibits seem quite normal. The big old firms like Cardeilhac hardly swallowed the new style – proof if any were needed of its limited appeal.

Elsewhere the seed was sown but did not take root. In Italy it was called the 'Stile Liberty' from the firm whence the imports came. Architecture was hardly affected, and silver not at all, if one is to judge from such works as *Liberty a Napoli* by Renato de Fusco, *L'eta di Liberty* by Italo Cremona, or the 1965 display at the Galleria Milano. The large and expanding firms, Calderoni, Bulgari and Buccellati, retained their static dignity in the face of all temptation. In the USA Louis Comfort Tiffany (1848–1943) with his original iridescent glass techniques and jewels in a style somewhere between Lalique and Henry Wilson, alternating surprisingly between the very elegant and the over-thick, was a lone phenomenon. He designed bronze and iron lamps and table decorations, but very little silver. Louis Sullivan got the message in Paris in 1874, as a student, but, alas, he never designed silver and his brilliant architecture with its lavish *art nouveau* details never caught on. The Americans preferred either heavy period reproductions or building based on economy of structure. This rough new society was not yet tired of artistic orthodoxy, nor yet settled enough for a *belle époque* high society to establish mannerisms of its own. Russia conversely was too tired. Peter Carl Fabergé (1846–1920) found sufficient novelty for his exquisite taste in the delicacies of eighteenth-century Dresden, and had no creative urge to unsettle him further. *Art nouveau* was relegated to the Moscow hotel buildings for the *nouveaux riches*.

1918–39

At the Bauhaus after 1918, the machine was at last recognized as a possible asset to art, not an inevitable enemy. 1890 to 1914 were the years of exploration when the status of artist-craftsmen dwindled; from 1918 to 1939 the machine conquered; since 1945 we positively like machines. A few exceptional artist-craftsmen are again strong, not, as Morris imagined, because they use their hands, but because of their brains and their creative originality.

So the period between the wars really was an in-between time for silver. Modern architecture, born in the previous decades, seemed inevitable. Adolf Loos, the Viennese architect, had caused a scandal with his essay of 1908 called *Ornament and Crime*. But when in 1930 he wrote 'I have liberated mankind from superfluous ornament', the public reaction may have been 'of course'. Many people assumed with indifference that the crafts would die and no new craft societies were formed, only groups for design. Yet the industrial designer was not yet hailed as the vital spark without which factories are damned. America was not yet quoted as the one great country which survives without artist-craftsmen, which treats its industrial designers properly. There were no great revolutionaries either in word or form.

The reason was social as much as artistic: the silver trade had not adjusted to the huge social upheaval of the 1914–18 war. The shops had to lower their sights to the middle classes, and those which refused to do so, like Garrards of London, the old Crown Jewelers, slowly went to the wall. Factories had to concentrate more on mass demand, and those who succeeded in mechanizing themselves in time, like Christofle of Paris, prospered enormously. Those who were caught by the 1929 slump, like William Hutton of Sheffield, went bankrupt.

In England, the Royal Academy 1935 exhibition of Industrial Art was intended to show the national achievement. The entrance balustrades in 'Staybrite' stainless steel, designed by Reco Capey, made by G. Johnson and given by Firth Vickers, still show how far we had got. Neither the fundamental truth and simplicity of the Bauhaus functional modern style, unsaleable as it proved to be, nor the necessity of pleasing the public, had impinged on the policy of British producers who mostly pursued an undistinguished mean. Nikolaus Pevsner has recorded his embarrassment at having to take foreign visitors round this show – a lack of conviction characterized British design, to remedy which the Design and Industries Association, father of today's Council of Industrial Design, had been formed as far back as May, 1915.

In silver, Harold Stabler was probably the best known name, partly because of his fine work for London Transport, Britain's finest design ambassador. Many of his pieces are now in the collection at Goldsmiths' Hall: the small step shapes, the tiny cubes and triangles typical of, say, Peter Behrens about 1910, coarsened under Stabler's touch, but still seem anaemic compared with the Frenchman, Puiforcat. Harry Murphy had the distinction of being Principal of the Central School of Arts and Crafts, unusual for a silversmith. Bernard Cuzner was the best known teacher in Birmingham. Perhaps most distinguished was R. M. Y. Gleadowe, art master at Winchester, and a fine draughtsman with a wonderful sense of line – another denial of the rule that only a silversmith can design silver sympathetically. Omar Ramsden was the best salesman of the time, if not the most sophisticated designer – hardly a cathedral exists without a cross of his, hardly a London guild without his cup. All these, and particularly the new young designers like Leslie Durbin, Cyril Shiner and R. G. Baxendale, owed some of their success to the medieval London guild, the Worshipful Company of Goldsmiths, whose reviving patronage now began to be a decisive influence led by G. R. Hughes.

Scandinavia led the silver world between the wars. Jensens in Denmark, now being followed by other older Danish firms to whom they had shown the light, continued to bring out new patterns; but public attention was by now firmly concentrated on the flamboyant early work of Georg Jensen himself, rather than the angular innovations of Nielsen and Bernadotte. It was in Sweden that the new style of bold straight lines and sharp points finally emerged. There was already a theory that a modern design should express the nature of the material used, for silver soft and curved, but the Swedes gave it the lie. Jacob Ängmann of GAB, Stockholm, Wiwen Nilsson with his own workshop in Lund, and Baron Erik Fleming of Atelier Borgila in Stockholm, were the leaders of their generation, and duly impressed many foreign visitors to the Stockholm 1930 Exhibition.

Elsewhere, silver was dogged by political and financial troubles. There were not many occasions for international comparisons. On the brink of war, in 1937, the Worshipful Company of Goldsmiths exhibited in Hitler's Berlin. Most of the modern work shown was remarkable for its elaborate ornament, more than for its invention. Emil Lettré of Berlin, Michael Wilm of Munich, Elisabeth Treskow of Cologne were the leaders, all reviving the old process of gold granulation. At the Louvre in 1966, there was a big show of Paris interior design of the time of 1925, *L'Art Déco*, with the Puiforcat family firm much in evidence. Jean Fouquet and Raymond Templier were two of the best established artist-craftsmen. It was more striking than sensitive, reminiscent of cinemas today. In Italy at the Galleria Milano, in 1965, there was a similar hark back to the great Paris international exhibition of decorative arts of 1925; *L'Art Déco* was undigested geometry.

1945–67

Cutlery is the big sensation of the past twenty years. Research into methods of production, of eating, of washing-up, and of laying tables, has continued as it did before the war, but the big change has been in costs. All over the industrial world labour shortage makes it difficult to polish by hand tens of millions of knives, forks and spoons; so simple shapes which can be polished by machine have swept the board. Furthermore, because of mass catering in big restaurants and canteens, cutlery when finished now gets much harder treatment than ever before; it gets bent and broken – if it is not stolen for souvenirs. Tough, cheap and sensible are the dominant requirements, and this tends to mean stainless steel, not EPNS or silver.

Viners, the big Sheffield cutlers, as late as 1960 made 85 per cent of their cutlery in EPNS; now in 1967 it is 75 per cent stainless steel. When the P & O Line built their magnificent *Oriana* for round the world service, they gave Robert Welch unprecedented power for a designer: he was to evolve and have made in stainless steel a complete outfit for the liner.

The history of spoons and forks covers no more than five centuries, but the changes in their use have been dramatic. In 1480, Jean Sulpice recorded that it was 'wrong to grab your food with both hands at once; meat should be taken with three fingers and too much should not be put into the mouth at the same time'. He also insisted that people should not scratch themselves at meals and then put their fingers in the food. Erasmus in *De civilitate morum puerilium*, published in 1530, gave a number of instructions: it was not good manners he said, 'to lick your greasy fingers or rub them on the jacket; they should be wiped on the napkin. Salt should be taken from the salt-cellar with the point of the knife, from which any grease has been removed by wiping it on the napkin or piece of bread'; it was a common joke that the mark of three fingers in the salt was the sign of a villain *(Tres digiti in salino impressi, vulgari joco, dicuntur agrestium insignia)*. It was also 'ridiculous to remove dirt from the shell of an egg with the finger-nails when it can be done more elegantly with a knife'.

In 1611 Thomas Coryat of Yeovil responded to the new implements: *I observed*, he says,

> *a custome in all those Italian Cities and Townes through which I passed, that is not used in any other country that I saw in my travels, neither doe I thinke that any other nation of Christendome doth use it, but only Italy. The Italian, and also most strangers that are commorant in Italy, doe alwaies at their meales use a little forke when they cut the meate; for while with their knife, which they hold in one hand, they cut the meate out of the dish, they fasten their forke which they hold in their other hande, upon the same dish, so that whatsoever he be that sitteth in the company of any others at meate, should unadvisedly touch the dish of meate with his fingers, from which all at the table doe cut he will give occasion of offence unto the company as having transgressed the lawes of good manners, insomuch for his error he shall be at least browbeaten, if not reprehended in words. This forme of feeding I understand is generally used in all places of Italy, their forkes being for the most part made of yron or steele, and some of silver, but those are used only be gentlemen. The reason of this their curiosity, is because the Italian cannot by any means endure to have his dish touched with fingers, seeing all men's fingers are not alike cleane.*

As late as the seventeenth century in France, kings still ate with their fingers; in Sweden forks were not in general use until the end of the nineteenth century. Even today national eating habits differ; Scandinavian cutlery looks like and is handled in the same way as surgical instruments; British cutlery is large and heavy, held firmly in each fist, and equal to the toughest meat and the roundest peas. In America one cuts up one's food with a knife, then puts down the knife and eats with the fork only, two separate processes, much as Europeans were doing in the sixteenth century.

Prices are bewildering in their variety; hand-forged modern silver cutlery by Gerald Benney in England for instance, sells at about £1000 for twelve place settings with fish knife and fork and six servers. At the other end of the scale is David Mellor's 'Minim' pattern in stainless steel, commissioned by the British Government to help civil servants to enjoy their food, produced by Walker & Hall Ltd, by the million, and costing £1 5s retail for a five-piece place setting. In-between is Robert Welch's 'Bistro' pattern with special heat-proofed rosewood handles, for Wiggin and Old Hall at £2 10s for two steak knives, two forks and two spoons. Gone are the days of comprehensive attention to the mouth; the Jensen 'Acorn' pattern consists of 226 pieces including melon knife and fork, caviar fork and beer can opener; whereas the new Magnus Stephensen 'Frégate' pattern is only twenty-eight pieces strong. English utility is reaching the hard-eating Danes.

The struggle to get established can be won in many ways; perhaps, like Louis Comfort Tiffany with his studio eighty years ago, or like Anthony Elson with Blunt and Wray church furnishers today, a designer may be lucky enough to start with financial assets of his own. Sometimes, like Arthur King of New York, he will find a wealthy patron to absorb some of the initial risk. More often he emerges from art school as a young designer, or leaves a big firm to set up independently, hoping that something will happen, that he will make a few sales and not stick. He may teach part-time, or earn a regular salary as consultant designer to a factory. But to succeed as a freelance silversmith he must have remarkable talent.

Competitions are invaluable in this context: they are the only means for new talent to break the established order. Sometimes there are money prizes, as with the Prix de la Ville de Genève, the most sought-after award in jewelry. The winner gets an amount rather greater than his work itself warrants – at Geneva in 1967 for an enamelled writing-set, for instance, he gets 5000 Swiss Francs, one of the four similar awards made there each year recently – and this splendid opportunity buoys up the hopes of competitors. It is the same sort of incentive as the race-track or the casino, but it has a larger professional importance. If one wins a big competition one's produce is automatically more desirable because one attracts more public attention.

In 1967 Roger King, the British jeweler, was exhibiting and selling his gold bracelets in Miami, Florida, under the legend 'creator of the jewel of the year', a reference to his victory in the British competition. The Topham Trophy competition each year provides an original piece for the winner of the second race on the famous Grand National course: the winner gets a £500 prize with a further £500 to make his piece. The London Goldsmiths', Silversmiths' and Jewelers' Art Council at Goldsmiths' Hall has each year since 1908 given prizes of up to £25 to specially talented craftsmen. From time to time, as with the Ascot Gold Cups given by the Queen at Ascot races in 1961, the Worshipful Company of Goldsmiths organizes national competitions for suitable patrons, usually offering a first prize of £500 or more. In 1967 the town of Pforzheim is celebrating its 200th anniversary as a jewelry centre with an international gold necklace competition, comparable to those held occasionally by the German Goldsmiths' Society, with a first prize of about £200.

Competitions sometimes work even without prizes. De Beers' annual Diamonds-International Awards scheme has no winner; instead, it commends designers of particular interest, whose work thus gets publicity and praise which may be worth more than money. Retailers seldom disclose the name of a designer or supplier from whom they buy work, but if they do, they like him to be famous, or at least to possess some glamorous tag, for instance to have won a great prize. Competitions ought to point to quality. Even if they fail to find it they draw attention to the existence of the ideal.

Most designers produce their best work when they are feeling inspired and confident, when in fact they have been commissioned by a good patron to do an interesting job. Competing with others may be useful in business or sport, but in art it often simply weakens resolve. Some of the best designers work only for one client, for fear that in a competition they will not distinguish themselves properly. A reputation once acquired must not be squandered in public. So a competition may not attract all the best talent available.

Some of the most sought-after awards are for mass-production prototypes: here the judges will select not a single phenomenon but a long-term commitment. The International Silver Company of the USA, for instance, celebrated its 100th birthday with a world-wide competition for cutlery, run by the New York Museum of Contemporary Crafts and won by the distinguished Swede Sven Arne Gillgren. Jensens of Copenhagen celebrated the 100th anniversary of Georg Jensen's birth with a Scandinavian competition in 1966; in 1954 they had marked the 50th year since Jensen opened his workshop, by their first such project, won by Tias Eckhoff, the daring young Norwegian. In 1967 Viners of Sheffield entered this world class with a first prize of 1000 guineas for cutlery, won by Robert Glover a student at Hornsey College of Art.

There may be several stages before the agony of decision passes. Judges will first scrutinize drawings: in Geneva in 1963 for instance there were 447 sketches from 94 sources, one third of whom were outside Switzerland. Then there may perhaps be invitations to half a dozen promising designs, resulting in models from which the winner is chosen. At the Liverpool Topham Trophy there are usually about 250 entries, at the New York Diamonds-International for jewels in 1966 there were about 1200. Only judges who are thought to combine impartiality with experience will inspire confidence and attract a good entry: for the 1967 Topham Trophy there were the director of the Walker Art Gallery, Hugh Scrutton; Graham Hughes, the author; silversmith Gerald Benney; and sculptor Robert Adams. For the New York Diamonds-International there were Karl Dittert, silversmith of Schwäbisch Gmünd, Paul Smith of the New York Museum of Contemporary Crafts, Emilio Pucci, the Florence couturier, and Ailsa Garland, Editor of the British *Woman's Journal*. Judges always try to produce a stimulating choice but they seldom succeed: the more character a design has, the less universal will be its appeal. Minority reports are unpopular, so the laurel often goes to the highest common factor, which may be only a quite boring second or third choice for each individual chooser.

Competitions provide one new and increasing source of patronage: companies and institutions are another. Unique among them is the Worshipful Company of Goldsmiths of the City of London, the only medieval guild in the world still vitally connected with its original craft, and still a pioneering influence in the cause of quality. Part medieval trade union, part parish community, part social fellowship, the Company has always been a society of individuals concerned with gold, silver and jewelry. Today, the Company – a City Livery Company – is a great charitable and educational foundation, but it still encourages high standards in its craft. The hall-mark (the word originates from the 'mark of Goldsmiths' Hall', London) still guarantees the purity of English gold and silver. At the annual Trial of the Pyx at Goldsmiths' Hall, the Company still tests the nation's coinage; it still 'binds' or enrols apprentices; it gives scholarships at suitable art schools; it arranges lectures and discussions at Goldsmiths' Hall and elsewhere; it stages exhibitions of good work at home and overseas; it has made films and filmstrips about the crafts, which are available on loan or for sale. It is not connected with any individual trade firm or association, but acts in the interests of the whole craft and industry with which it has always been closely associated. It has a proud history, but respect for the past does not mean neglect for the future; Britain's designers, encouraged by the Company, have never been in better shape.

The Company owns the largest collection of modern works in the country, to which additions are continually being made; this collection is the only one of its type in the world. Here are some of the best recent silversmithing products. Both large and small pieces are included: they give encouraging proof of the revival of this craft which is steadily gaining strength. The selection of work by living craftsmen at Goldsmiths' Hall is exhibited very widely; in 1966 the Goldsmiths' pieces were shown in San Francisco, Los Angeles, New Orleans, Minneapolis, Seattle, Boston, Montreal, Toronto, Miami, Lyons, Helsinki, Stuttgart, Hong Kong and Leeds.

Museums like the dead; living designers are too unpredictable, their reputations too insecure, their work too expensive. Many museums are now buying *art nouveau* silver which, at the time it was made, they mostly thought outrageous and disgusting. In London the Victoria and Albert, for instance, cold-shouldered Ashbee during his life but now buys his work and places it on show; whereas the pieces by designers living today who win Design Centre awards, like David Mellor or Robert Welch, are kept in storage, not exhibited. The museums of Stockholm, Copenhagen, Helsinki, Bergen and Oslo all buy new work regularly and advertise and exhibit it. In the LGA Museum, Stuttgart, there is the fine display of ordinary household utensils, with the best modern cutlery representation in the world. The beautiful Pforzheim Reuchlinhaus now buys more jewels than plate, but has close and admirable links with its local industry. In Sheffield in the City Museum is the world's best old cutlery, and at the Klingenmuseum (blade museum) in Solingen is a more modern group. In the New York Museum of Modern Art are some choice examples both of *art nouveau* handwork and of modern mass-production. But new silver has, sadly, not yet really impinged on art history. No public display is complete without its antiques, but although there is more variety today than ever before in new silver and cutlery, there is less attention from the museums. New painting and sculpture are considered acceptable by them, but new silver is negligible.

Interest in modern design is not a dominant British characteristic. It was therefore gratifying to find the London House of Lords discussing it in June 1964 as a subject worthy of national attention. What is good design, they asked, and the answer, at least for Lord Conesford, was honesty. He deplored the sort of dishonesty which makes an electric fire look like flickering coal, or which makes plastic imitate wood instead of being itself, or, most absurd, which makes the product of one perfectly worthy manufacturing process disguise itself as another. We all know the type of thin metal bowl first spun to a neat shape on a lathe, then hand hammered with feeble dimples which spoil the shape and catch the dirt. Bad deceptions are certainly revolting.

But art is illusion. One wonders what Bernini or Pozzo, Cuvillies or Zimmermann would call their own marvellous dreams. What of Mantegna's painted ceiling balustrade at Mantua, or Rembrandt's conspirators in Stockholm, with swords seemingly diagonal to the canvas, or Wren's upper nave wall at St Paul's, supporting nothing, sheltering nobody, put there simply to look nice from the outside. Indeed what of the stone volutes throughout all architecture, which don't bear an honest load of weight? These deceptions are good, and here is the crux of the matter.

It's conviction that counts, whether in painting or silver or industrial design. Good design is impossible to define except in terms of the artist's instinct; if a good artist believes in a design, it will probably succeed, even if it is a deception. This point of view allows for inspiration and originality, and implies that design is an imaginative art, not a mathematical problem. Design is good if it conveys the artist's conception with confidence and strength.

'Good design', however, is normally used in the very limited sense of honesty and practicability, and in this context the phrase is worn out. As the British *Journal of the Society of Industrial Artists* recently said, 'that helpless, meaningless, spiritless cliché "good design" needs a fundamental life-saving operation to give it reality again. As a concept it is as dead and obsolete an expression in the design world as is the dodo in the animal world.' We are all discovering that industrial design, if it is simply sensible, can be simply boring, even though it is still preferable to the gimmicky frills of twenty years ago.

There has been a remarkable recent outburst of energy in enamels. In England Stefan Knapp, with his magnificent exhibitions in the Hanover Galleries, and his huge outside murals at London Airport, and then at Seagram Building, New York, has rightly become famous; similarly, Marit Aschan, with regular exhibitions in the art galleries of London, New York and Oslo, is turning from painting to enamelling, and proving herself the meeting ground between fine and applied art. In Sweden enamel murals of a gay simplicity are quite a common feature of the best modern architecture. There is Lage Lindell's 40-foot long mural at Västerås Town Hall (1960) and for an office block in Örebro (1961), or the exotic Stig Lindberg 3-foot bar decorations for the new ship *Svea Jarl* (1962). The Gustavsberg china factory, founded in 1827, expanded when it was bought by the Co-operative Society in 1937, and pioneered these enamels from 1949, under their staff instructor Eje Öberg. In France, the monks of Ligugé, in the medieval tradition, are nourishing a specialized craft. Sigurd Persson worked there; more recently they have made champlevé enamel work by Georges Braque, and the notable signed Georges Rouault crucifix, 55 × 38 cms, for the Chapel of the Messageries Maritimes steamer, *Viet-Nam* (1952). The Metarco Gallery in New York has just opened with the aim of selling new enamels and is showing work by Americans such as Karp or Maxwell Evans. All these enterprises, with Burch-Korrodi in Zürich, point to a great future: the strict technical demands of enamel provide a healthy discipline for exuberant creative artists.

Modern silver has come to mean a luxury product made in small workshops; the big factories mostly make very few modern designs, concentrating on cheap methods of reproducing traditional shapes mostly in base metal. For instance, Viners of Sheffield, who stamp out huge quantities of copper dishes, electroplate them with silver, and roll on to the surface with steel-embossed rollers, elaborate eighteenth-century type scroll work. WMF at Geislingen or Christofle in Paris tell the same story – the public yearns for the antique appearance. So new designs are made in small quantities by people who really care about them, some of whom may design a new pattern for mass-production once every two or three years.

Sigurd Persson is typical of the post-war generation in Sweden, with Barbro Littmark of Bolin, and Lars Fleming of Borgila; this is the land of hard restraint. In Denmark, outside the Jensen team, there is Karen Strand, chiefly jeweler, Hansen and Hingelberg, Cohr, Dragsted and Michelsen. Style here is softer, but it is still too impersonal for most of the world markets. Small workshops in Germany are too numerous to list, but the City of Düsseldorf is typical of the country of the economic miracle: excellent new retail shops there, each managed by a young designer, mainly of jewels, are: Hartkopf, Kern, Krall, Kutzer, Münstermann, Schmitt & Staib, Wels, and Weyersberg. René Kern and Weyersberg, much the largest, are also the least advanced artistically. Germany is the home either of a heavy outmoded functionalism, or of the delightful modern whimsy of such young artists as Hermann Jünger. Surprisingly, modern silver seems hardly to have reached France and Italy, though Enrico Sabbatini has done some impressive light pieces for mass-production by Krupp of Milan, and sculptors like Bruno Martinazzi occasionally create a silver sensation. But silver has always been a product of the cold North rather than the hot South.

In England, the economics of this industry have given their clearest verdict. The big factories cannot raise enough capital to mechanize completely, and with the existing competition from small outworkers, they cannot afford the necessary high wages to compete with new undertakings in steel and motor-cars. So the big factories are insecure, and in no mood to dabble in art. It is in small workshops like those of David Mellor, Robert Welch, Louis Osman, or Gerald Benney, each employing five or fifteen people, that new enterprise is found. This is the country for commissioned work   towns and universities, companies and clubs order pieces specially designed for them, just as architects may order sculpture or painting. Personality is stronger in British silver than anywhere else in the world. This seems to be the pattern for the future: stainless steel and base metal from the big factories for convenience, silver from the small men for beauty.

237 Teapot *c.* 1880
   D Christopher Dresser
   MO James Dixon Ltd, Sheffield
   The idea of functional design
   pioneered by Dresser in a brilliant
and precocious manner, gave way to *art nouveau* and did not become fashionable until the 1930s in Scandinavia

237

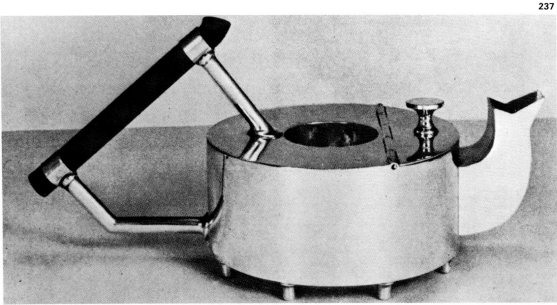

238

239

238 The Ascot Gold Cup 1884. The sinuous lines of *art nouveau* are already appearing. Sold at Christies, London for £1900 in May 1965. Height 19½ in.
  D Sir Alfred Gilbert

239 Tazza 1897
  DM Gilbert Marks, one of the strongest and most sensible silversmiths of the British arts and crafts revival group. This movement generally and consciously revived original design, but, as here, it hardly accepted *art nouveau* although the style was invented by its members
  O Worshipful Company of Goldsmiths

240 Fish knife and fork in silver plated nickel
  D *c.* 1900 Charles Rennie Mackintosh either for himself or for William Davidson whose house 'Windyhill' at Kilmacolm he also designed
  O Museum of Modern Art, New York

240

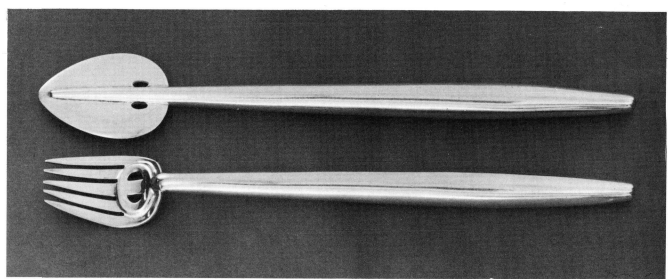

241

241 **D** Charles Rennie Mackintosh
1904

242

242 **D** Charles Rennie Mackintosh
1905 for Miss Cranston's
Glasgow tea rooms
O Glasgow School of Art

144

243 Box in pewter and lead partly
   silver plated
   D Perhaps Charles Rennie
      Mackintosh *c.* 1900
   O Alan Irvine

244 Mustard pot
   D C. R. Ashbee
   M Guild of Handicraft before
      1900
   O Kunstgewerbemuseum,
      Zürich

245 Silver and green glass decanter
   D C. R. Ashbee
   M Guild of Handicraft 1904
   O Victoria and Albert Museum,
      London

244/245

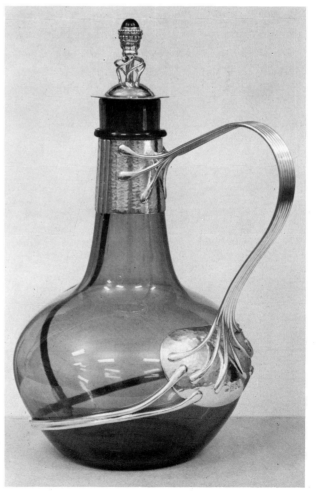

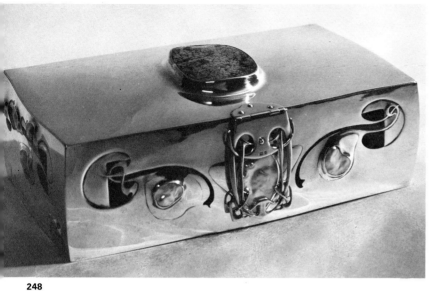

248

249

250

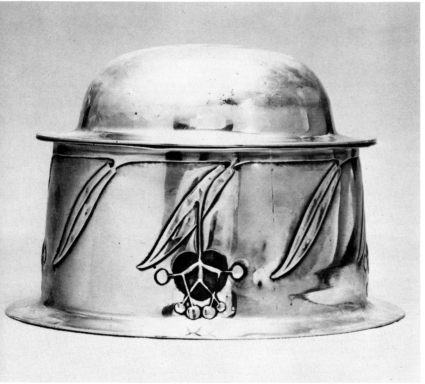

251

246 Silver and nacre coffee set
   M William Hutton and Sons Ltd of Sheffield for Goldsmiths and Silversmiths Co. Ltd of London, shown at the Paris exhibition 1900, where Huttons were the only British firm to win a prize
   O Kunstindustrimuseet, Oslo

247 Covered bowl
   D C. R. Ashbee
   M Guild of Handicraft *c.* 1903
   O Kunstgewerbemuseum, Zürich. There are very similar bowls at the Victoria and Albert Museum, in the Shirley Bury Collection London, and in the Darmstadt Museum

248 Jewel casket: silver mother of pearl turquoise enamel
   D Alexander Knox
   M William Craythorne for Liberty of London *c.* 1900
   O Museum of Modern Art, New York. Length 11¼ in.

249 Cigarette box
   M William Hutton 1901
   O Worshipful Company of Goldsmiths

250 Silver and enamel box
   DM W. H. Haseler & Co., Birmingham 1903 for Liberty
   O Victoria and Albert Museum, London

251 Pewter and blue enamel inkstand
   M Liberty *c.* 1911
   O Museum of Modern Art, New York. Height 3⅝ in.

252 **DM** W. H. Haseler, Birmingham
1904 for Liberty
O Victoria and Albert Museum,
London

253 Cup: ivory enamel and stones, in
champlevé enamel letters 'Drink
and fear not your man'
DM Harold Stabler *c.* 1910
O Victoria and Albert Museum,
London

256

255

254/256 Furnishings designed 1897–1908 by Henry Wilson for the huge St Bartholomew's Church, Brighton, are the grandest English example of *art nouveau*, even though they are incomplete. The church, designed by Edmund Scott for the keen philanthropist Father Arthur Douglas Wagnor, was built 1872–4. It is higher than Westminster Abbey or Amiens Cathedral, and was intended as a spiritual warning to holidaymakers leaving the nearby railway station. Henry Wilson designed the chalice (254) in 1898 and the baldacchino in 1899, the altar rail (255) *c.* 1905

257

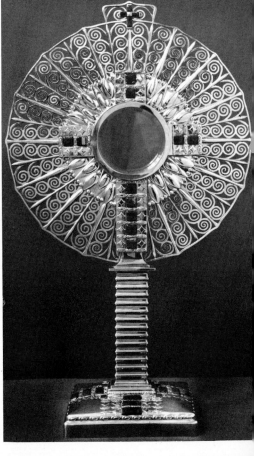

258

257/260 The St Leopold am Steinhof church, Vienna. This was the winning design in a competition 1902. The architect Otto Wagner was one of the creators of the Viennese *Sezession*. The church, built 1904—7, is probably the grandest example of his work, with details by him; Kolo Moser and the Wiener Werkstätte

261 Silver malachite four rows 'pearl'
drops
   D C. O. Czeschka
   M Wiener Werkstätte *c*. 1910
   O Museum für Kunsthandwerk,
   Frankfurt

262 **D** Joseph Olbrich *c*. 1910
   O Hessisches Landesmuseum,
   Darmstadt

263 Pewter candelabrum
   D Joseph Olbrich 1901
   O Museum of Modern Art,
   New York

264 Girandole for nine candles:
hammered and cast silver
   D Hans Bolek
   M Alfred Pollak 1912
   O Österreichisches Museum für
   angewandte Kunst. Vienna

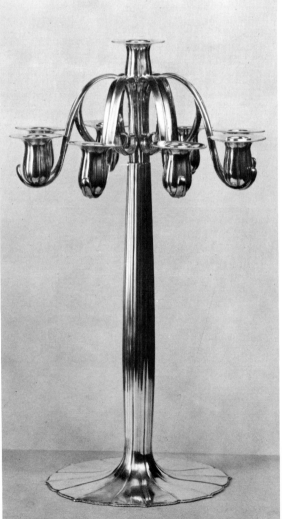

265 **D** Joseph Hoffmann
M Wiener Werkstätte *c.* 1910
O Österreichisches Museum für
angewandte Kunst, Vienna

266 Pewter jug
D Theodor Pogačnik *c.* 1900
O Wiener Kunstgewerbeschule

267 Table setting: silver with gilt
inside. Diam. 26 cm.
D Eduard Wimmer 1910 for
Friedmann, Vienna
O Österreichisches Museum für
angewandte Kunst, Vienna

268 Vase, silver: base metal insert,
daffodil shape
D Joseph Hoffmann
M Wiener Werkstätte 1912
O Österreichisches Museum für
angewandte Kunst, Vienna

265/266
267/268

153

269

270

269 Flower bowl
   D Joseph Hoffmann 1909
   O LGA Museum, Stuttgart

270 Covered cup: silver and lapis
   lazuli
   D Kolo Moser 1908
   O Österreichisches Museum für
   angewandte Kunst, Vienna

271 Tea kettle
   D Henri Van de Velde *c.* 1904
   O Kunstgewerbemuseum,
   Zürich. A similar set is in the
   Stadtmuseum, Munich

272 Tea kettle
   D Van de Velde *c.* 1902
   O Kunstgewerbemuseum,
   Zürich

273 Teaset: boxwood and silver
   D Van de Velde *c.* 1905
   O Kunstgewerbemuseum,
   Zürich

271

272

273

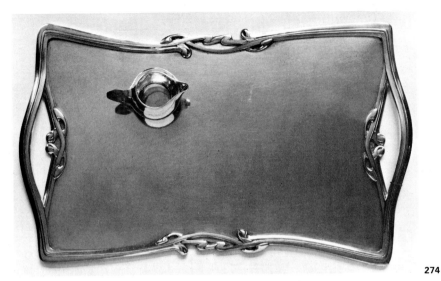

274 Tea tray
  D Van de Velde *c.* 1902
  O Kunstgewerbemuseum,
    Zürich

275 Three bowls
  D Van de Velde *c.* 1905
  O Kunstgewerbemuseum,
    Zürich

276 Ivory and silver
  D Van de Velde 1922
  O Kunstgewerbemuseum,
    Zürich

274

275

276

277

277 Bowl
   D Van de Velde in Weimar
      c. 1906
   O LGA Museum, Stuttgart

278 **D** Van de Velde 1900
   O Musées Royaux d'Art et
      d'Histoire, Brussels.
      Ht 23 in.

279 Silver plated bronze
   D Van de Velde in Weimar
      c. 1900. Ht 28 cm.
   O LGA Museum, Stuttgart

278

279

280 **D** Van de Velde 1902–3
O Kunstgewerbemuseum, Zürich. A similar set is in the Oslo Museum, the Osthaus Museum Hagen, and in the Hamburg Museum für Kunst und Gewerbe. Van de Velde's early silver was mostly made by the Weimar court jewelers Hans and Wilhelm Müller, and by his own pupil Albert Feinauer: Van de Velde says in his autobiography, published in Munich in 1962, these were among the best things he ever did. Fishknife 20·5 cm.
M Theodor Müller, Weimar

281 **D** Van de Velde *c.* 1902
O Kunstgewerbemuseum, Zürich
282 **D** H. P. Berlage 1912
M W. Voet, Haarlem
O Kröller-Müller Museum, Otterlo

283

284

285

283 **D** Professor Christiansen
   M P. Brückmann, Heilbronn 1902
   O Bodo Glaub collection, Cologne

284 **M** P. Brückmann, Heilbronn 1915
   O LGA Museum, Stuttgart

285 Silver. Dessert knife 24 cm., butter knife 20·5 cm.
   D Richard Riemerschmid, Munich 1900
   O LGA Museum, Stuttgart

286 Silver
   D Peter Behrens
   M M. J. Rückert, Mainz *c.* 1900
   O Hessisches Landesmuseum, Darmstadt

287 Silver, designed for the exhibition of the artists' colony in Darmstadt 'Ein Dokument Deutscher Kunst' 1901
   D Peter Behrens
   M M. J. Rückert, Mainz *c.* 1900
   O Hessisches Landesmuseum, Darmstadt

286

287

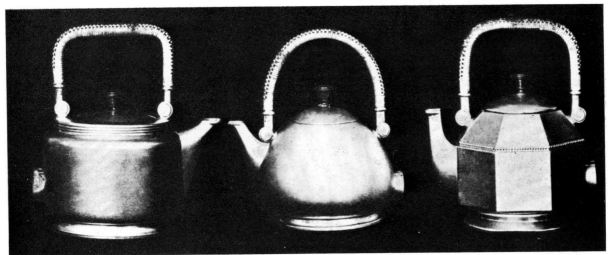

291

292

288 Pioneering industrial design:
three electric kettles, base metal
D Peter Behrens 1910–12
M AEG one of Germany's
biggest companies, Berlin

289 **D** Richard Riemerschmid 1914

290 Silver. Dated 1900. Ht 23 cm.
M M. Mau. Court Jeweler,
Dresden
O Museum für Kunsthandwerk,
Frankfurt

291 Silver beaker, gilt inside
Ht 10 cm.
D Peter Behrens, Dresden 1899
O LGA Museum, Stuttgart

292 Beaker
D Peter Behrens, Munich 1907
O LGA Museum, Stuttgart

161

293

294

295

293 Covered cup: silver with cast
   engraved and textured surface
   DM Ernst Riegel, Darmstadt 1906
       Ht 47·3 cm.
   O LGA Museum, Stuttgart

294 Moss agate body, in the form of
   a duck, silver lid (with silver
   comb), with amethysts, carnelians
   (the twig in the duck's beak),
   copper and gold. Ebony base
   with silver, mother-of-pearl and
   amethysts
   DM Ernst Riegel, Darmstadt 1911
   O Museum für Kunsthandwerk,
       Frankfurt

295 Pewter jug
   D Albin Müller, architect and
       designer, from 1906 professor
       in the Darmstadt technical
       school 1871–1941
   M E. Hueck, Lüdenscheid
       c. 1910. Ht 34·5 cm.
   O Hessisches Landesmuseum,
       Darmstadt

296

297

296 Pewter teaset
D Hugo Leven, sculptor and designer active in Düsseldorf and Bremen with workshop in Paris, designer for the distinguished pewter series 'Kayser-Zinn' 1874–1956
M J. P. Kayser Sohn, Krefeld 1899. Ht 14·5 cm.
O Hessisches Landesmuseum, Darmstadt. Exhibited in Turin and Darmstadt 1902

297 Pewter teapot
D Hugo Leven
M J. P. Kayser Sohn, marked 'Kayser-Zinn' c. 1900 Ht 21·5 cm.
O Museum für Kunsthandwerk, Frankfurt

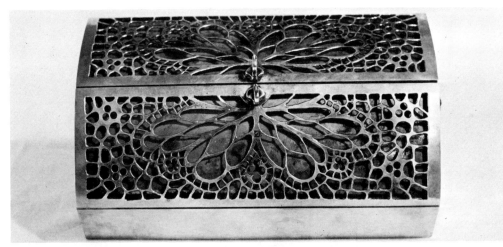

301

302

298 Sugar basin: silver and ivory
     Chocolate jug: silver and wood
     M Cardeilhac, Paris, *c*. 1895
           Hts 18·5 cm. and 21 cm.
     O Kunstindustrimuseet, Oslo

299 Silver box for communion bread
     DM Eric Ehrström 1907
     O Tampere Cathedral, Finland

300 Typical of the sophisticated,
     delicate and unpractical enamels
     which at this time gained a world
     reputation for Norway
     D Gustav Gaudernack
     M David Andersen, Oslo 1901
     O Museum für Kunsthandwerk,
           Frankfurt

301 Jewel casket: ivory and silver
     DM Philippe Wolfers,
            Brussels 1899–1902
            Length 16½ in.
     O Musées Royaux d'Art et
            d'Histoire, Brussels

302 Silver communion service
     DM Eric Ehrström 1907
     O Tampere Cathedral, Finland

303 Vase: silver gilt and enamel
D Torolf Prytz
M Tostrup *c.* 1900
Ht 22 cm.
O Kunstindustrimuseet, Oslo

304 Bowl: filigree and enamel
DM Gustav Gaudernack for
David Andersen 1907
Diam 16·5 cm.
O Kunstindustrimuseet, Oslo

305 **D** Johan Sirnes
M David Andersen *c.* 1916

306 **D** Johan Sirnes
M David Andersen *c.* 1916

305

306

307

309

308

307 Silver beakers
   D Johan Rohde
   M A. Dragsted 1913. Beaker on
       right. Ht 22 cm.
   O Kunstindustrimuseet,
       Copenhagen

308 Coffee pot
   D Just Andersen, Copenhagen
   M Ballin's successor P. Hertz
       1914. Ht 19·5 cm.
   O Kunstindustrimuseet,
       Copenhagen

309 The simple and sensible
      Scandinavian modern tradition
      is already beginning with this
      silver cutlery
   D Knud Engelhardt
   M Carl Cohr, Fredericia 1908
   O Kunstindustrimuseet,
       Copenhagen

310 Tea box
   D Thorvald Bindesbøll
   M A. Michelsen *c.* 1900.
       Ht 8 cm.
   O Kunstindustrimuseet,
       Copenhagen

311 Chocolate bowl
   D Thorvald Bindesbøll
   M A. Michelsen 1900
       Diam 15·5 cm.
   O Kunstindustrimuseet,
       Copenhagen

312 Beaker
   D Thorvald Bindesbøll
   M A. Michelsen 1908
       Ht 19·5 cm.
   O Kunstindustrimuseet,
       Copenhagen

313 **D** Thorvald Bindesbøll
   M A. Michelsen 1909
       Ht 13 cm.
   O Kunstindustrimuseet,
       Copenhagen

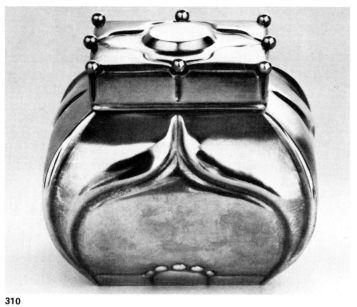

310

311

312

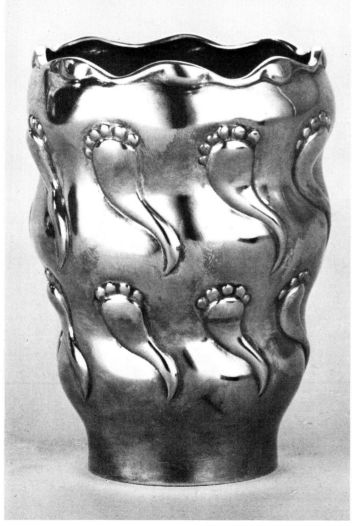

313

314 **315**

314 Large silver cup and cover
DM Omar Ramsden, London
1916
Collection the late Queen Mary

315 **DM** Omar Ramsden 1929
O The Worshipful Company of
Goldsmiths. Ht 5¾ in.

316 Pocket cigarette case with gold
coat of arms in the characteristic
English manner, by tradition made
by the artist for his own use
throughout his life
DM Omar Ramsden 1919
O The Worshipful Company of
Goldsmiths. Length 4½ in.

316

317

318

319

317 With silver, pearls and many
colours of enamel, this is one of
the richest, best-made, and most
satisfying products of the whole
British arts and crafts movement,
whose often quaint and badly-
made designs were by then
disciplined by fine craftsmanship
and practical requirements.
Casket given by the citizens of
Birmingham to HRH the Princess
Royal on her marriage to Lord
Lascelles 1922
DM Arthur Gaskin, Birmingham

318 Silver and blue enamel cigar box
given by the Worshipful Company
of Goldsmiths to Sir Robert
Williams Bt, for 50 years a
member of their Court 1884–1934
DM Harold Stabler, London 1935
O Sir David Williams Bt

319 Typical of the geometric and
functional style of the 1930s,
this silver bowl with blue enamel
rim was commissioned by the
Worshipful Company of
Goldsmiths in 1933. Diam. $4\frac{1}{4}$ in.
D Jane Barnard
M Edward Barnard & Sons Ltd,
London

171

320

321

322

320 Silver chased alms dish with
Romanesque inspiration, typica
of the best church work which
by then was growing out of th
Gothic revival
D Eric Gill
M H. G. Murphy and E. B.
Wilson, London 1930
O The Worshipful Company of
Goldsmiths

321 Silver engraved alms dish.
Although Gleadowe was much
less famous than Gill, his silver
was more important. He evolved
a personal style of exquisite
linear work, related to his own
enthusiasm for drawing at
Winchester College
D R. M. Y. Gleadowe
M H. G. Murphy 1930.
Engraver: George Friend
O The Worshipful Company of
Goldsmiths

322 Large silver dish and beaker,
with chased spiral pattern.
Before knives and forks became
known, *c.* 1700, large dishes
were necessary on grand tables,
for washing the fingers, and they
often contained perfumed water
in which rose petals floated,
hence the name rosewater dishes.
The custom of decorating tables
with these fine bowls lives on
today for our visual delight
D R. M. Y. Gleadowe
M Wakely & Wheeler Ltd 1938
O Corpus Christi College,
Cambridge. Given by
Professor A. L. Goodhart,
Fellow 1920–6

323 Altar plate, given to Washington
Cathedral DC by HM King
George VI of England to
commemorate Anglo-American
friendship during the war
D R. Y. Goodden, after a design
by his uncle R. M. Y.
Gleadowe at the Savoy
Chapel, London of 1937
M Blunt & Wray 1952

324 Teaset, with die-struck stepped
ornament
D Harold Stabler, London, 1928
M Wakely & Wheeler Ltd for the
Goldsmiths & Silversmiths
Company Ltd
Jug Ht 8¼ in.
O The Worshipful Company of
Goldsmiths

323

324

325

325 Hot water jug: nickel silver with
ebony and raffia
D Marianne Brandt
M Bauhaus 1924
O Museum of Modern Art,
New York. Given by the
designer

326 Hand-made silver cutlery
DM Wenzel Hablic 1918—25
O Museum für Kunft und
Gerwerbe, Hamburg

327 Candelabra box and menu
holder
D Gio Ponti *c.* 1927

328 Vanity case: silver and lacquer
D Raymond Templier, Paris 1930

329 Nickel silver and ebony
D Marianne Brandt
M Bauhaus 1924
O Museum of Modern Art,
New York

326

327

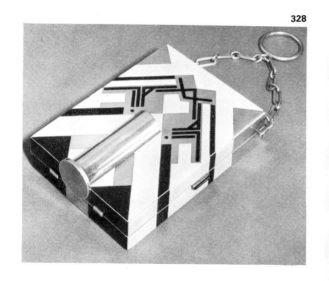

328

329

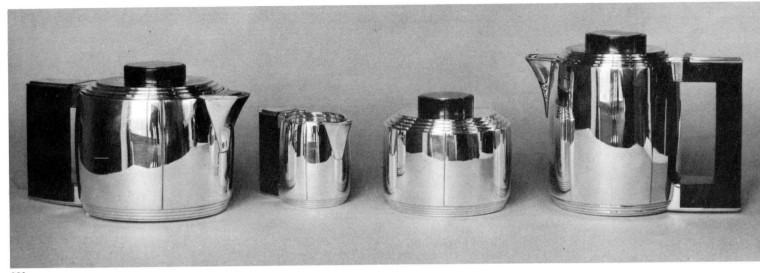

330

330 **DM** Jean Puiforcat, Paris *c*. 1925

331 **DM** Jean Puiforcat, Paris *c*. 1925

331

332

333

334

332 **DM** Jean Puiforcat, Paris *c.* 1925

333 **DM** Jean Puiforcat, Paris *c.* 1925

334 Probably Europe's most popular
flatware design for a period of
over a century. 'Baguette',
however, never caught on in
England and Scandinavia,
probably because it was
unfashionably large. Produced by
Christofle first in 1861, and still
being sold by them in 1967 at
the rate of 100,000 pieces
annually, 'Baguette' was made
by several other factories
including Hutton of Sheffield,
whose ambitious, expensive and
sophisticated machine tooling
for export of this pattern was
frustrated by the 1929 slump,
and therefore partly caused the
firm's dissolution
DM Christofle, Paris *c.* 1920

335 Communion set
  D Henry Ericsson 1931
  O Paul Church, Helsinki,
  Finland

336 Communion set
  D Gunilla Jung 1935
  O Church of Mikael Agricola,
  Helsinki, Finland

338

337 Wine flagon
DM Wiwen Nilsson, Lund,
Sweden

338 Altar set: gold, silver and
Colombian mahogany
DM Wiwen Nilsson, Lund,
Sweden 1964
O Lund Cathedral

179

339

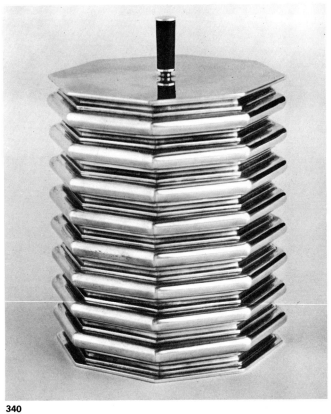

340

339 Bonbonnière Sweet Box
   D Kay Fisker
   M A. Michelsen 1926
   O Kunstindustrimuseet,
      Copenhagen

340 Tobacco jar
   D Kay Bojesen
   M A. Michelsen 1928
   O Kunstindustrimuseet,
      Copenhagen

341 Saucepan
   DM Inger Møller 1937. Ht 13 cm.
   O Kunstindustrimuseet,
      Copenhagen

342 **D** Magnus Stephensen
   M Kay Bojesen 1938
   O Kunstindustrimuseet,
      Copenhagen

342

341

180

343

343 Silver and enamel candle
holders. 24 ct gold inlay
DM Frederick A. Miller,
Cleveland Art Institute,
USA 1964. Ht 4½ in.

344 Brass silver plated teapot
DM Fred Fenster, Madison,
Wisconsin, USA 1966

344

345 Torah shield. Jewish ritual plate
is perhaps more conservative
even than Christian church plate,
but both could be, as they were
in the Middle Ages, the main
source of patronage for craftsmen
DM Bernard Bernstein, Bronx,
New York, USA 1964

346 Chalice: silver-gilt and ivory.
Gained for Herr Gebhart the
State prize of Westphalia
DM Friedrich Gebhart, Roxel
Nr. Münster, Germany 1966.
Ht 17 cm.

347 Bronze chalice with silver-gilt
cup and rim inserted, silver-gilt
paten
DM Olaf Skoogfors,
Philadelphia USA 1965

348 Jewish altar. One of the thirty-
seven prize winners from fifteen
countries at the 1966
Internationales Kunsthandwerk
exhibition at the LGA Museum,
Stuttgart. 'Jiskor' Mannah and
Memorial table, tin and pewter.
Lead welding technique. The
piece is a memorial to the six
million Jews of Europe

murdered by the Nazis: above
symbols of prison, confusion and
death, appears the sacred seven-
armed Jewish candelabrum.
DM Kurt Pfefferman, Haifa,
Israel 1966. Pfefferman was
born Berlin 1919, emigrated
to Palestine 1939, won a
Bavarian state prize gold
medal Munich 1962.

349 Silver gilt and rock crystal
monstrance
DM Friedrich Becker,
Düsseldorf 1963
O St Heinrich Church, Marl

350

351

352

353

184

354

355

350 A late and distinguished example
of the hard functional style
DM Andreas Moritz, Nuremberg
1954

351 Chalice
DM Paul Günther Hartkopf,
Düsseldorf 1964

352 Silver bowl
DM Max Zehrer, Tittmoning,
Germany 1960

353 Large church candlesticks
DM Paul Günther Hartkopf,
Düsseldorf 1965

354 Goblet: stainless steel with red
synthetic corundum balls (ruby),
gilt inside. The designer has tried
successfully to find an honest
use for synthetic stones
DM Friedrich Becker,
Düsseldorf 1966

355 Bell: bronze with rock crystal
ball
DM Friedrich Becker 1963
O City of Düsseldorf

356 Silver teaset
   DM Max Fröhlich, Zürich 1961

357 Hospital communion set, six
   goblets
   DM Max Fröhlich, Zürich 1953

358 Candelabrum, given by the Zürich
   goldsmiths as thanks for the
   Worshipful Company's
   exhibitions during the British
   Trade Fair, Zürich. Engraved
   'Given 1963 to the Worshipful
   Company of Goldsmiths of
   London from the Association of
   Zürich Goldsmiths united in the
   effort for art and craft'
   D G. Heinz
   M Meister, Zürich
   O Worshipful Company of
   Goldsmiths

359 Gold cup with patchwork of
   square enamels in red and blue,
   interior enamel in turquoise.
   The Prize of the City of Geneva
   is probably the world's most
   popular award of its type. Three
   sections are normally devoted to
   watches and jewelry and one to
   enamels, for which the 1966
   winner is shown here
   D Luigino Vignando
   M Brauchi & Fils, La Chaux-
      de-Fonds, Switzerland

358

359

187

360

360 Silver ice bowl
DM Bertel Gardberg, Helsinki
1956

361 Silver teapot, black wood handle
DM Bertel Gardberg 1960

362 Silver coffee pot, rosewood
handle
DM Bertel Gardberg 1958

361

362

363 Silver communion flagon, gold
details
DM Bertel Gardberg 1965
O Tapiola Church, Espoo

364 Silver flagon, gold details
DM Bertel Gardberg 1966.
Ht 28 cm. Prize winner at the
Internationales
Kunsthandwerk exhibition
1966 at the LGA Museum,
Stuttgart

365 Silver coffee service
D Tapio Wirkkala, Helsinki 1959
M Kahvikalusto Hopeaa

363

364

365

366 Casket commissioned by British
Cellophane Ltd to celebrate the
Festival of Britain 1951. The
details engraved in silver and
applied in gold, suggest the
treatment of fibre involved in
cellophane production
D R. G. Baxendale
M Mappin & Webb Ltd,
Sheffield 1952

367 Large rose-water dish: silver
parcel gilt. Commissioned by the
Rubber Trade Association of
London to celebrate the Festival
of Britain 1951
D R. G. Baxendale
M Mappin & Webb Ltd,
Sheffield 1951

368 Large dish given by Queen
Elizabeth the Queen Mother to
the University College of
Rhodesia and Nyasaland on her
installation as President
DM Leslie Durbin 1957

367

368

369

371

369 The Queen's Cup. The fiery
ball suggests festivities; the royal
arms carved inside the foot are
particularly fine. Commissioned
by the Worshipful Company of
Goldsmiths to celebrate the
coronation of HM Queen
Elizabeth II 1953. Ht 18 in.
D Professor R. Y. Goodden
M Wakely & Wheeler Ltd.
Engraver T. C. F. Wise

370 Silver wine mug with gold stem
given by the Worshipful
Company of Goldsmiths to
Sir Owen Wansbrough-Jones
on his joining their Court
DM Leslie Durbin 1957. Ht. 5 in.

371 Silver and yew, given to Downing
College Chapel, Cambridge by
H. Darlow as a war memorial.
D Professor R. Y. Goodden
M Leslie Durbin 1953. Ht 28 in.

372

373

374

372 Badge in gold and precious stones, commissioned by the Smithsonian Institution, Washington to celebrate the bi-centenary of the birth of its founder in 1765, the illegitimate son of the Duke of Northumberland
DM Leslie Durbin, London 1966

373 Small christening bowl given by Graham Hughes to his godson Andrew Armytage
DM Eric Clements, Birmingham 1953

374 Silver with black leather handles, each embellished with a cast and engraved silver tea leaf. Commissioned by the Tea Centre, London to celebrate the coronation of HM Queen Elizabeth II 1953
D Eric Clements
M Wakely & Wheeler Ltd 1954

375 One of three silver cups commissioned by Feeney Charitable Trust for Birmingham Art Gallery to celebrate the coronation 1953. The sun on the finial represents the Sheffield plate mark of Matthew Boulton the great Birmingham 18th-century industrial pioneer who invented mass-production for consumer goods
D Eric Clements
M Nayler Bros. for Payne of Oxford 1953

376 Detail of the largest of the three cups; on the side is the royal arms with those of Birmingham and of Feeney; on the finial above the flames is St Elegius, patron of European goldsmiths, martyred by burning in Birmingham; on the knop bulls symbolize the Birmingham Bull Ring
D Eric Clements
M Nayler Bros. for Payne of Oxford 1954. Engraver T. C. F. Wise

377 Given to TM the King and Queen of Denmark by HM Queen Elizabeth II on her state visit to Copenhagen in May 1957
D Eric Clements
M Wakely & Wheeler Ltd. Engraver T. C. F. Wise

378 Inkstand
D Alex Styles. Length 15 in.
M Wakely & Wheeler Ltd 1965
O Runcorn Urban District Council, Cheshire, England, given by the first woman chairman

375

376

377

378

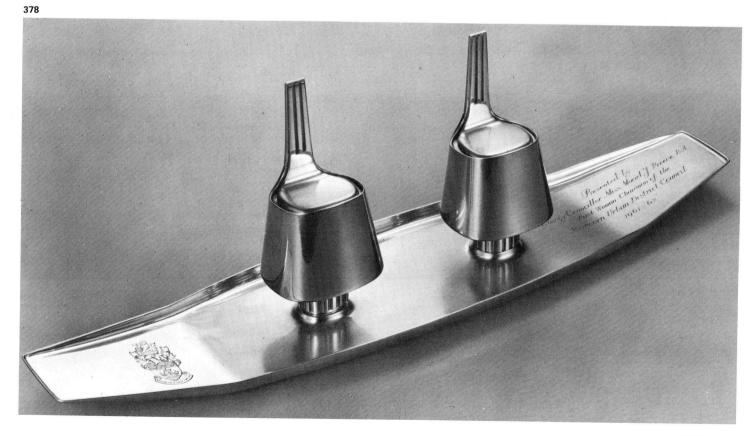

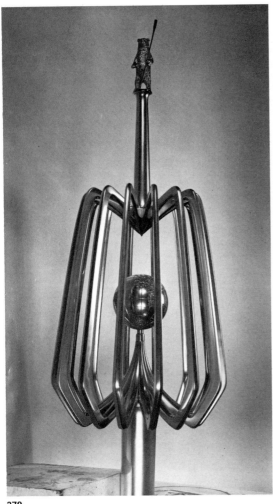

379

379 Mace commissioned by Cornell
University, New York
D Eric Clements
M Wakely & Wheeler Ltd,
London 1965

380 Silver gilt paper weights
DM Keith Tyssen, Sheffield, as
a basic design exercise with
his students at the College
of Art in their Diploma in Art
and Design course 1965
O Worshipful Company of
Goldsmiths. Ht 1 in.

380

381

382

381 One of 2 candelabra given by
the Worshipful Company of
Goldsmiths to the University of
Exeter
DM Keith Tyssen 1966. Ht 23 in.

382 Centrepiece with hors-d'œuvre
dishes; second prize in the
Topham Trophy competition,
Liverpool 1964
DM Keith Tyssen. Centre length
21 in.
O Worshipful Company of
Goldsmiths

383 Trophy: perspex sphere with
silver centre rhodium plated to
avoid tarnish, and silver and
black slate base. Entry in the
Topham Trophy competition,
Liverpool 1965
DM Peter Wheeler, London
O The Worshipful Company of
Goldsmiths. Ht 6 in.

383

384 Given by the Worshipful Company
of Goldsmiths to Timothy Dwight
College, Yale University, USA
in memory of John Marshall
Phillips, a Liveryman of the
Company 1941–55 and Fellow
of the College 1956. Ht 12½ in.
DM David Mellor, Sheffield

385 'Pride' pattern teapot made in
silver and in EPNS. Design
Centre Award 1959, Council of
Industrial Design, London
D David Mellor. Ht 5½ in.
M Walker & Hall Ltd, Sheffield
O The Worshipful Company of
Goldsmiths

386 Silver cigarette boxes with gilt
textured lids
DM David Mellor 1965
O The Worshipful Company of
Goldsmiths. Length 3¾ in.

387/8 Part of a set of modern table
plate called 'Embassy'
commissioned by the British
Government for use in British
embassies, the first being
Warsaw. Design Centre Award,
London 1965. Teapot Ht 6¾ in.
DM David Mellor
O The Worshipful Company of
Goldsmiths

387

388

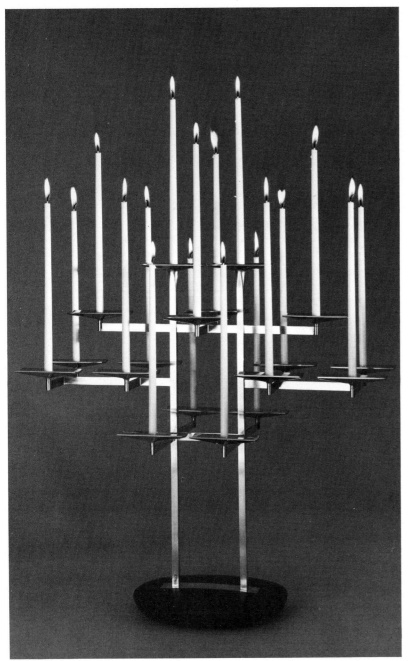

389

390

389 Candelabrum given by Sheffield
Design Council to the City of
Sheffield. Ht 21 in. Width 23 in.
D David Mellor
M David Mellor and Sheffield
College of Art 1960

390 Seven-light candelabrum, a
prototype for Jewish ritual use
DM Robert Welch, Chipping
Campden 1958. Ht 14½ in.
O The Worshipful Company of
Goldsmiths

198

391

392

393

394

391 Part of a large range of modern
domestic plate commissioned by
Heals department store, London
DM Robert Welch 1966. Ht 5 in.
O The Worshipful Company of
Goldsmiths

392 Coffee pot, nylon rim and handle
DM Gerald Whiles, Birmingham
1962. Ht 4¾ in.
O The Worshipful Company of
Goldsmiths

393 Teaset with ivory rims
D Robert Welch 1962
M Wakely & Wheeler Ltd
O The Worshipful Company of
Goldsmiths. Ht 5½ in.

394 Large triangular water jug,
2½ pt capacity. Part of a group
given by the Worshipful Company
of Goldsmiths to the new
University of Sussex
D Gerald Whiles. Ht 11 in.
M Silver Workshop Ltd 1966

395

395 Large coffee pot commissioned
by the Grenadier Guards
D Keith Redfern
M Silver Workshop Ltd 1966

396 Part of a gift by the Worshipful
Company of Goldsmiths to the
new University of York. An
ingenious use of mass-produced
tubes and plates with
elaborately etched and textured
gilt coats of arms
D Atholl Hill, London 1965
M Wakely & Wheeler Ltd

397 Large bowl with gilt arms, as 396
D Atholl Hill 1965
M Wakely & Wheeler Ltd

398 Gilt cigarette boxes, an original
product of hand-operated lathes
and machine tools
DM Stuart Devlin, London
O The Worshipful Company of
Goldsmiths. Ht $3\frac{1}{2}$ in.

399 Mace: the flanges stove enamelled
black, and gilt. Given by the
Worshipful Company of
Goldsmiths to the new University
of Bath. Length 38 in.
DM Stuart Devlin 1966

396

397

398

399

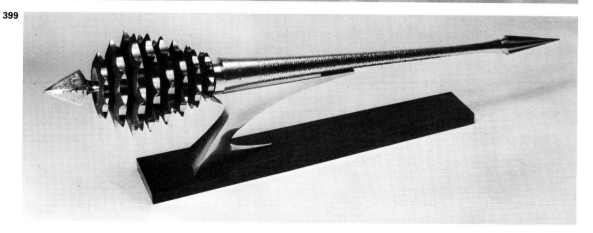

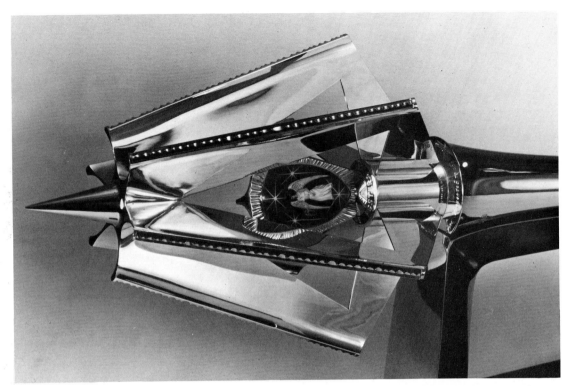

400

400 Mace, given to the University of
Melbourne, Australia, by Colonel
Aubrey Gibson
DM Stuart Devlin 1965

401 Candelabrum, partly gilt. Given
to the new City University,
London, by the Association of
Past Students of the College and
the present students in the
College in May 1966
DM Stuart Devlin. Diam. 13 in.

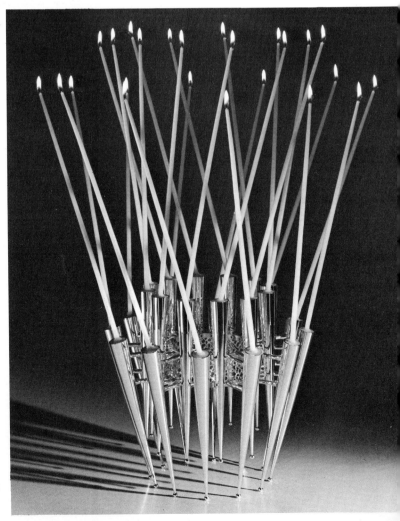

401

402 Mace in stainless steel and
enamel given to the new
University of Bradford by the
four Yorkshire universities: York,
Sheffield, Leeds, and Hull 1967
DM Desmond Clen-Murphy,
Brighton

403 Large bowl given by De Beers
Consolidated Mines Ltd to the
Guldsmedsskolan (trade
technical school), near Linköping,
Sweden, to celebrate its 60th
Jubilee
DM Desmond Clen-Murphy,
Brighton 1966

403

404

404 Stainless steel cross and
candlesticks
DM Brian Asquith. Ht 4 ft 6 in.
O St Augustine's Church,
Basford, Nottingham

405 Stainless steel candlesticks
DM Brian Asquith, Sheffield.
Ht 1 ft
O St Mary's Church, Hampton,
Middlesex

405

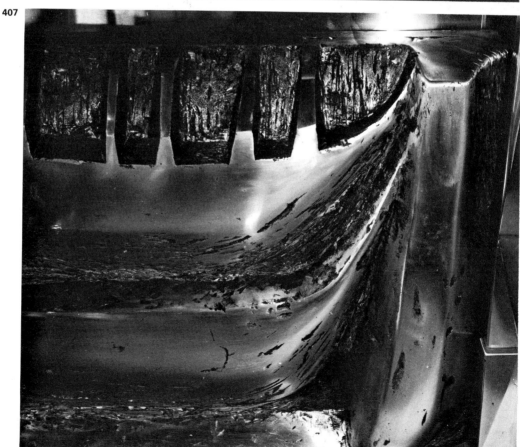

406 Bronze sculpture 'Elements –
Fire – Steel' suggesting processes
of steel production, Sheffield's
primary industry. Detail
DM Brian Asquith
O Westminster Bank, Sheffield

407 Detail of 406

408

409

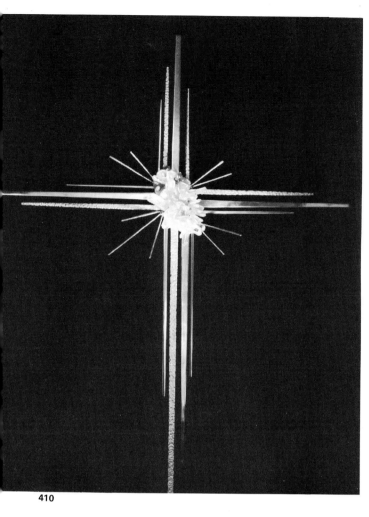

408 Crucifix: bronzed steel
 DM Robert Adams, London 1964
  Ht 24 in. Width 28 in.
 O St Louis Priory, Missouri, USA

409 Detail of table ornament in steel, deep etched to a knife edge, then re-etched. Perhaps a suggestion of the ornamental style of the future, reminiscent of drawings by Paul Klee.
 DM Jussuf Hussein Abbo, London 1967

410 Altar cross with illuminated rock crystal centre
 DM John Donald, London 1963. Ht 4 ft. Width 3 ft
 O Birmingham Anglican Cathedral

411 Large alms dish, silver with gilt letters 'alpha' and 'omega' incised, cast and hand-raised with 14lb sledge hammer
 D Louis Osman, Rotherfield
 M Louis Osman, The Morris Singer Co. Ltd and Desmond Clen-Murphy 1959
 O The Worshipful Company of Goldsmiths

412 Rock crystal foot of a large altar set
 DM Louis Osman 1959
 O The Worshipful Company of Goldsmiths

410

411

412

413 The Ely Cross, one of the very few 20th-century Christian statements showing real conviction and originality, and, therefore, unusually disturbing; the crucifix in gold, the first important commissioned sculpture modelled by Graham Sutherland; the heart, in black nielloed silver, symbolizes continuing life. The cross is silver with square silver nielloed slabs as hands, and gold fingers symbolizing Our Lord's authority. The whole is made from 45 separate parts and stands 42 in. high. Commissioned by Ely Cathedral and the Worshipful Company of Goldsmiths, rejected by the Cathedral and now in the Roman Catholic chapel of Mr Emery Reves at Villa La Pausa, Roquebrune.

D Louis Osman and Graham Sutherland

M Louis Osman, Johnson Matthey Ltd, Morris Singer Ltd, and Desmond Clen-Murphy 1964

See *Connoisseur*, October 1964

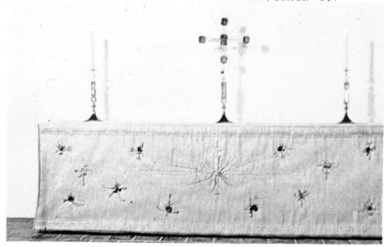

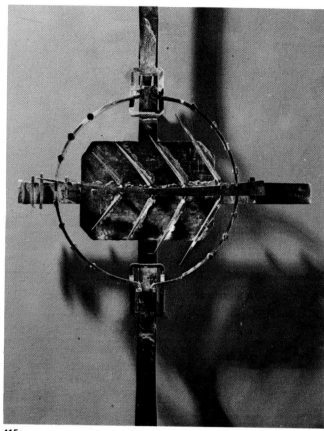

414   415

414 Altar (10 ft long) with frontal representing the stones from the door of the Temple and the New Jerusalem in the Book of Revelation; green jasper, lapis lazuli, dioptase, rock crystal, red and white onyx, red jasper, yellow agate, beryl, yellow serpentine, green chalcedony, sapphire, amethyst.

D Louis Osman

M Louis Osman, Thea Somérleatte (embroiderer) and E. Wolfe and Co. (jewelers)

O St James Major Church, Shere, Surrey

415 Silver and perspex altar and silver cross at King's College, London, Theological Hostel Chapel, Vincent Square. A Cross of Despair and Hope balanced. Cruelty and Evil cannot trap mankind because of the eternal power of the Cross. The spiked circle could symbolize a crown of thorns of endurance, humility and justice, or a temporal crown of ambition and worldly power. The palm frond holds the spring of the trap (which is made to work) as Christ's victory over sin and death

DM Louis Osman 1967

O*

416 Detail of 414. Lapis lazuli **D** Geoffrey Clarke.

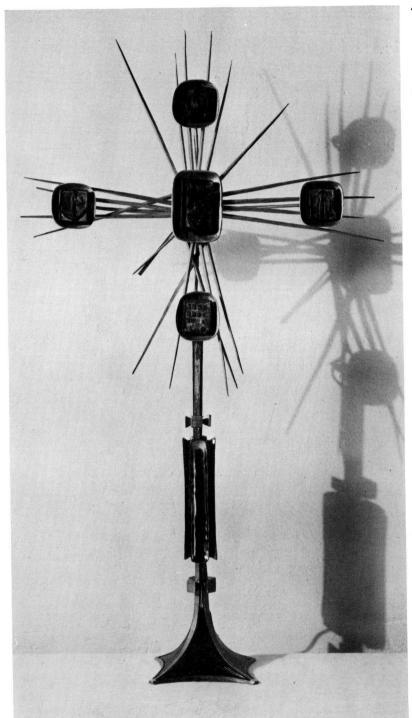

417

417 Cross of gilt iron, bronze, wood
and red leather
D Louis Osman
M Louis Osman and Geoffrey
Clarke 1959
O Shere Church, Surrey

418 Nickel bronze cross with crystal
slabs inset in boxes and lights
inside. Given by the architect
Sir Basil Spence to Coventry
Cathedral
DM Geoffrey Clarke, Stowmarket,
Suffolk 1958. Ht 7 ft
O Coventry Cathedral. The
Chapel of the Cross (West
Crypt)

418

419

421

419 Bronze low relief 'Christ in our
Humanity'. New church
furnishings showing such genuine
intensity of feeling are extremely
rare. One of the series made
1947–66, shown at the Welz
Gallery, Salzburg 1966
70 × 50 cm
DM Giacomo Manzù, Milan.
Self-trained sculptor after
carpenter's apprenticeship, *b.*
Bergamo 1908

420 Deposition with cardinal, another
low relief from the series 'Christ
in our Humanity'.

421 'Door of Death' bronze at St
Peter's Rome, details of the
lower left-hand side. *Top left:*
'Death of Abel'. *Top right:* 'Death
of Joseph'. *Bottom left:* 'Death by
ordeal'. *Bottom right:* 'Death of
Pope John XXIII'
DM Manzù 1964

422 Decanter and goblets in case, one of the richest designs in recent decades, given to Sir Leslie Gamage to celebrate his 70th birthday by his friends in the General Electric Company Ltd of which he was Chairman. The falcon suggests Sir Leslie flying round the world, and the heraldry refers to the schools, regiments and institutions of which he was a member

D Louis Osman
M Gerald Benney, Louis Osman, David Wynne and others 1957

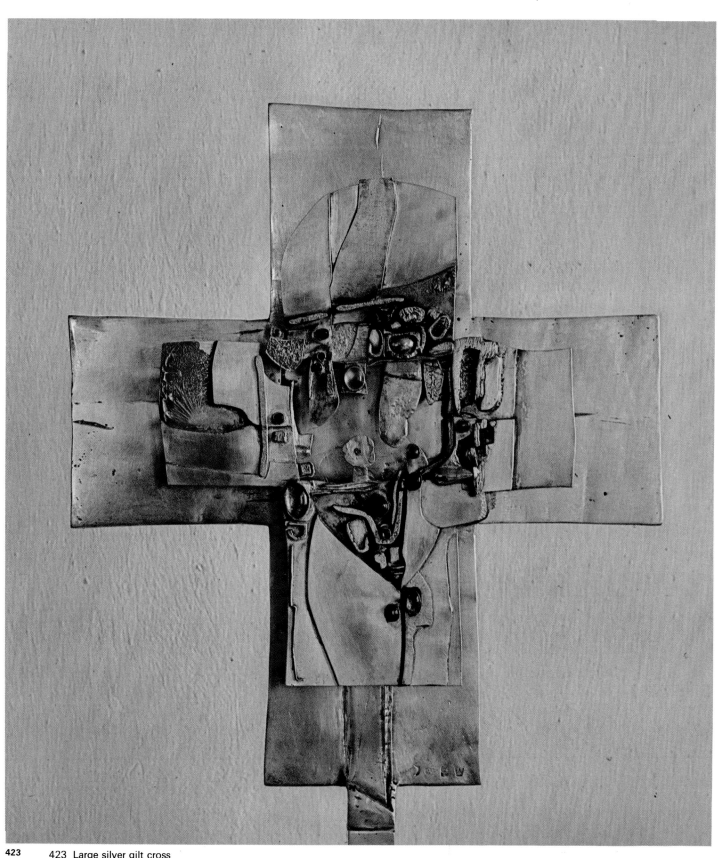

423 Large silver gilt cross
commissioned by the German
government to show in the
Montreal World Fair, Expo 67
DM Hermann Jünger,
Taufkirchen, Nr Munich 1967

424 Silver and EPNS 'Pride' cutlery,
first produced as a prototype at
the Royal College of Art London
1954, subsequently made in
quantity. Now one of the firm's
best sellers. Design Centre
Award London 1957
D David Mellor, Sheffield
M Walker & Hall Ltd (now
British Silverware Ltd)
O The Victoria and Albert
Museum and the Worshipful
Company of Goldsmiths

425 'Embassy' made in sterling silver
for use in British embassies, the
first being Warsaw. First
produced 1963. Design Centre
Award London 1965
DM David Mellor and Fletcher
cutlers, Sheffield
O Victoria and Albert Museum,
the Worshipful Company of
Goldsmiths

426 'Dimension', a typical high-quality
modern American design in
sterling silver of impersonal
character, intended thereby to
have a wide appeal
DM Reed & Barton

427 Stainless steel prototype, the
winning design in the
international competition for
cutlery organized by Viners of
Sheffield 1967. Chosen from
nearly 300 flatware designs from
27 countries, and given a first
prize of 1,000 guineas
DM Robert Martin Glover,
Hornsey College of Art,
London

428 Silver and ivory handles
DM Stuart Devlin, London 1960
O The Worshipful Company of
Goldsmiths

429 Silver cutlery, hand forged, gilt
handles
DM Stuart Devlin, London 1966
O Dr Arnold P. Gold, Harrison,
New York

430 Stainless steel cutlery, 'Vogue',
possibly the first modern design
ever to be produced in Australia
D Stuart Devlin 1967
M Wilshire Cutlery Company,
Melbourne

429

430

431

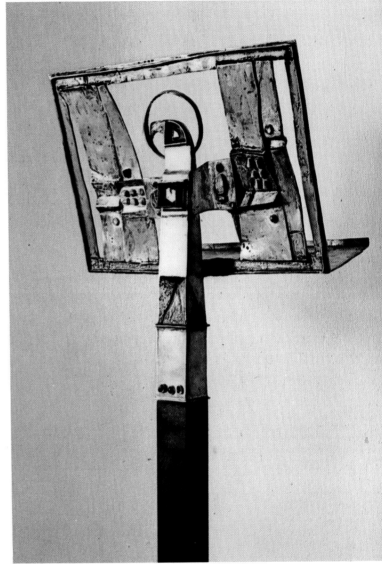

432

431 Chalice for a Roman Catholic
   priest in Augsburg. Jünger's
   invention of an almost Byzantine
   richness in an entirely modern
   idiom, is one of the inspiring
   features of post-war German
   silver and jewelry
   DM Hermann Jünger 1967

432 Silver cast lectern in the form of
   an eagle for the chapel of the
   Evangelic Academy in the castle
   of Tutzing, Starnbergersee
   DM Hermann Jünger 1967

433

434

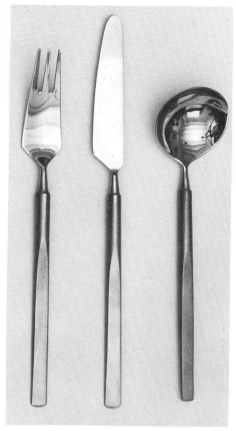

435

436

437

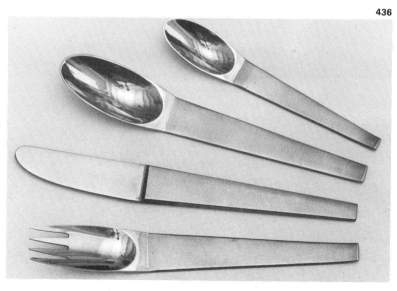

433 **D** Gio Ponti 1952
    M Krupp, Milan 1956

434 **M** Neuzeughammer
        Ambosswerk 1961
    O LGA Museum, Stuttgart

435 Stainless steel
    D Eric Herløw
    M Universal Steel Company,
        Copenhagen 1954
    O LGA Museum, Stuttgart

436 **D** Carl Auböck, Vienna
    M Neuzeughammer
        Ambosswerk 1957
    O LGA Museum, Stuttgart

437 Stainless steel, intended partly
    for use in the SAS Royal Hotel,
    Copenhagen; the whole building,
    furniture and fittings were
    designed by the famous Danish
    architect Arne Jacobsen. This
    fine design by Jacobsen proved
    unpopular there and was taken
    out of use but has since become
    a good international seller
    M A. Michelsen, Copenhagen
        1957

    O The Worshipful Company of
        Goldsmiths; Victoria and
        Albert Museum, London;
        Kunstindustrimuseum,
        Copenhagen; Nordenfjeldske
        Museum, Trondheim;
        Stedelijk Museum,
        Amsterdam.

438

439

440

441

438 **D** Tapio Wirkkala, Helsinki 1958
    M Christofle, Paris

439 Silver
    D Tapio Wirkkala, Helsinki 1956

440 Silver knife and fork
    D Tapio Wirkkala, Helsinki.
    One of five preliminary
    awards in the competition
    for sterling silver flatware
    organized by the Museum of
    Contemporary Crafts, New
    York, for the International
    Silver Company 1960

441 Silver knife, fork and spoon
    'Silverwing' or 'Duo' (in France)
    D Tapio Wirkkala 1954
    M Hopeakeskus, Hämeenlinna,
    Finland; and Christofle, Paris

Biographies

AUTHOR'S NOTE
*I have concentrated on giving long biographies of the more prominent firms
and designers, particularly those who specialize in silver. Many short
notes on jewelry firms, and artist/jewelers are to be found in the corresponding
section of my companion book* Modern Jewelry.   G.H.

Adam, Francis 1878–1961
Britain

Learnt craft in Hungary in wrought iron. Started in London with Starkie Gardner making wrought iron gates for Holyrood and other famous places. After 1920 started own workshop where he carried out staircase for the Tate Gallery, gold plate for Truro Cathedral, wrought iron for St Paul's and finally gold and silverwork of the Stalingrad Sword (by his pupil Leslie Durbin). Taught at Central School of Arts and Crafts, London, 1906–58.

Adams, Robert 1917–
Britain

Sculptor. Studio workshop Highgate, London. Studied Northampton. Designed a few jewels from 1958, and series for Goldsmiths' Hall 1961 international exhibition, made by HJ Company. Won two prizes in De Beers 1961 jewelry competition at Goldsmiths' Hall. Large architectural reliefs at Eltham Comprehensive School, Sekers showroom Sloane Street London, Gelsenkirchen Opera House Germany, Queen's Gardens Hull, Custom House London Airport 1967, and liners *Canberra* and *Transvaal Castle*. Bronze crucifix in St Louis priory, Missouri, USA, in which several international artists worked. Exhibits regularly at Gimpel Gallery London. Plate 408, page 140

Adie Brothers
Britain, Birmingham

Founded 1879 as Adie and Lovekin. 1907 registered Birmingham Assay Office as Adie Bros. During the period between the wars Percy Adie led the firm, leading Birmingham manufacturers; later controlled by Tony Adie (1909–65), Frank Taylor, and now British Silverware Ltd. Pages 74 76

Andersen, David
Norway

Retail and manufacturing firm, Carl Johansgate, Oslo, established 1876 by David Andersen. Present directors are his son, grandson and great grandson. Many exhibitions and prizes. See Tostrup, Gaudernack. Plates 300 304–6

Andersen, Just 1884–1943
Denmark

Firm established by Just Andersen (b Greenland) in 1915 in Copenhagen. Awarded Grand Prix Barcelona 1829, Brussels 1935, Paris 1937. Many exhibitions; included in 'The arts of Denmark', Metropolitan Museum, New York 1960, and other cities in USA. Plates 150, 308, page 97

Ängman, Jacob 1876–1942
See GAB pages 30–31 137, plates 43–52

Antrobus, Philip Limited
Britain

6 Old Bond Street, London, W1. Retail and manufacturing firm. Established by Philip Antrobus in 1815 in a Birmingham yard making silver spurs and horse harness trappings. Later with his son John started fine jewels. About 1900 Richard Leycester Antrobus (whose son now directs firm) opened workshop in London, Burlington Place. Fire destroyed premises in Hedon Street, but Richard and his son William, fifth generation, battled through. William, the recent head, who died 1945, used to go out with suitcase of gems and if he made a £25 sale in a day he was doing well. Widow still director at 82, with Leslie Ricketts, who began as apprentice at 15 in 1934 for 5s a week, less 1s 2d insurance; he won more design competitions than any other apprentice of his time. Charlie Hatch, designer for thirty-seven years, guesses he has designed over 10,000 rings. Until war, workshops in Manchester and London; also retail showrooms in Regent Place from 1930. Moved to present address in 1949.

Arström, Folke Emanuel 1907–

Studied at Linköping, and Stockholm technical school and Royal Academy of Arts (Konsthögskolan). Taught painting for 16 years in apprentice and trade schools in Stockholm. Paintings and graphic design, including Viking maps for the State Historical Museum 1934, and arms of the provinces of Sweden, and lettered addresses of honour to many distinguished people including Toscanini. Since 1931 concentrated on metal and plastics design. Many exhibitions including pewter in Nordiska Company 1933; silver, stainless steel and royal addresses at the National Museum, 1940, 1944. Silver medal of the Handicraft Association 1927; Milan Triennale gold silver and bronze medals 1951–60; California State Fair gold medals 1955–60; 1961 Gregor Paulsson statuette. Represented in many museums. Head designer for Gense, 1940–64. Plates 59 61, page 38.

Aschan, Marit 1919–

Jeweler and enameller, own workshop London. Has exhibited for last seven years at the Van Diemen-Lilienfeld Galleries, New York, and Oslo. 3rd one-man show Leicester Gallery, London 1967. See *Connoisseur*, April 1966.

Ashbee, Charles Robert 1863–1842, Britain

Architect, designer and writer. Educated Wellington and King's College, Cambridge. Articled to G. F. Bodley, architect, 1883, living and taking evening classes at Toynbee Hall, the Cambridge University settlement in the East End. 1888 founded Guild of Handicraft, acting as director and supervising designer. Concentrated on all aspects of interior design, including furniture and silverware. Influential in improvement of factory workshops, and advocated use of the machine in applied art as a necessity, though handwork more desirable. 1902 moved Guild of Handicraft to Chipping Campden, Gloucestershire, where old houses were restored and agricultural work undertaken, retaining a London shop. Guild wound up 1908. Author of *Silverwork and Jewellery* 1912; 1917 Master of the Art Workers' Guild; 1963 Centenary exhibition, Victoria & Albert Museum. Plates 244–5 247, pages 10 71 132–3, 140.

Asprey & Company Limited, Britain

Bond Street, London. Retail and manufacturing jewelers and silversmiths. Founded Mitcham 1781 by William Asprey, descendant of French Huguenot family. 1808 he and his son Charles established as dressing-case fitters; 1830s Charles opened shop at 49 New Bond Street; 1848 moved to no. 165. Honours at Great Exhibition 1851; Gold Medal, Paris 1855, 1862. Principals now Philip, Eric, Algernon, Harry Asprey. Sixth generation of Aspreys is rising in the firm. Leather, gold, jewelry factory above retail shop. Since 1954 active in Middle East. 1959 acquired Birch and Gaydon as branch in City of London. See page 73.

Asquith, Brian 1930–, Britain

Studied sculpture Sheffield College of Art and Royal College of Art 1947–51. Design consultant with wide range including window furniture, saws and kettles; silversmith. Three Design Centre awards, 1961, '62 and '65. Now pioneering use of cast stainless steel for tableware. Plates 404–7, page 76

Atkin Brothers, Britain, leading Sheffield firm.

Founded by Thomas Law *c.* 1738. Truro works of *c.* 1850 bought Bradbury 1947, absorbed by Adie and C. J. Vander 1958. Page 75

Auld, J. Leslie 1914–
Scotland

Silversmith, trained at the Belfast School of Art and Royal College of Art, London. Subsequently head of the silversmithing department Glasgow School of Art. Designed a King's Gold Vase for Ascot and dish given by London City Livery Companies to New York to commemorate the New York World Fair 1939–40.

Axelsson, Ainar 1920–
Sweden

Studied Stockholm and NAAS. Worked as silversmith in Holland, taught there. 1949 first prize in a competition for office façade. Much church plate. Staff designer at GAB, deputy of Gillgren.

Barkentin & Krall
Britain

Ies Barkentin b *c.* 1800 Denmark d 1881 Regent Street, London. Registered hall-mark at Goldsmiths' Hall 1861; became jeweler to Danish Princess Alexandra on her arrival 1863. Carl Christof Krall, a Czech, joined him *c.* 1868 but did not register till 1881. Made the Milton *Paradise Lost* shield in Westminster Abbey. Ies's son George Slater Barkentin (b 1840) went to USA as illustrator of American Civil War for *Illustrated London News*. Married an American. Krall became head of London firm, employing M. Cecil Oliver, later famous as freelance letterer. George's son, William Slater Barkentin (1874–1962) ran the firm Barkentin Studios, 45 West 18th Street, New York, and became a well-known artist in USA. Married Marjorie 1908. Designed Coca Cola sign *c.* 1910 and she said, 'It goes before me all the days of my life.' Had three sons, one grand-daughter. London firm ended in 1932, with Walter Stoye as managing director and chief designer. Famous for ecclesiastical metalwork and responsible for Giles Gilbert Scott's Lady chapel cross and candlesticks at Liverpool Cathedral (1909), for Wanamaker gold and silver altar plate and altars at Sandringham Church, used by Royal Family, and for Wanamaker gifts to St Mark's church Philadelphia (1904–9) where Wanamaker store is. See page 131

Barker Brothers (so called from *c.* 1860)
Britain, Birmingham

Manufacturers of 'Silver and electroplated and Britannia metalwares, spoons and forks' founded *c.* 1820 by partners Barker and Creed, *c.* 1864 moved to Unity Works, Paradise Street. Elkingtons were biggest Birmingham factory, Barker's second, and third, to whom some of their early dies passed, were Greenbergh,

Great Hampton Street, under their dynamic manager Mr Greenbergh; they later became Ellis & Co. William and Matthias Barker inherited Barker business from their father, the founder. William wrote to Matthias on 2 March 1864:

I accept your proposal for salary – with this arrangement – that you shall as a rule be at the Works by 8.10 a.m. – and stay in the Evening when Orders have to be sent away – of importance – rather than this should be neglected I would prefer paying the Expense of a Cab to enable you to be punctual in the morning. I would rather you take occasional time in the middle of the day to attend to any private important matters and I am also anxious for all the Workpeople to be at the Works at their proper time that it may enable the Warehouse men to get their orders off at a more reasonable time in the Evening and so close at 8 o'clock wherever it is possible. Also I would suggest that you receive £50 per Quarter and the balance at the end of each year.

William's sons, Frank and Herbert, in 1907 formed a Public Company, moving to present factory in Unity Works, Constitution Hill; the two brothers disagreed as to their fair shares and the newspapers criticized proposals, the *Birmingham Post* noting 'Barker Bros. have traded nearly a hundred years, but as a business proposition their Company will not stand a serious scrutiny.' An amusing instance of journalists proving wrong. At that time, average annual profit was about £9000. By 1914 some six hundred employees. In 1921, amalgamation with Levi and Salaman, whose factory next to Birmingham assay office in New Hall Street, specialized in 'toilet goods' such as enamel and tortoiseshell brush sets; and with Potosi Silver Company, spoon and fork makers whose trade mark was an eagle. Joint managing directors were Frank Barker and C. J. Levi. New York showroom opened *c.* 1920, then as now, only permanent showroom established by any British silver manufacturer in the USA. 1931 purchased Ellis & Co (trade mark, a pineapple, with trefoil clover used for nickel silver and EPNS). Frank Barker died in 1942 being succeeded by Martin Levi. In 1963 Ellis factory was moved into Barker's, business since 1931 called Barker-Ellis, the country's leading producers of elaborate traditional silver-plated copper, still with substantial production of varied sterling silver; commonest Barker trade mark is a lion rampant (particularly in Canada). 1964 Martin Levi retired, became President; fresh team of directors, no longer connected with original families. 1964 profit £21,000, fell to £14,000 in 1965. £25,000 spent on new machinery, the whole firm now employing some 360 people, a significant indication of falling profits and contracting labour force in this trade, against which greater mechanization is necessary. Own some Matthew Boulton dies.

Barnard, Edward & Sons Limited
Britain, London

Manufacturing silversmiths. This firm has perhaps longest continuous history and descends from largest number of famous silversmiths of any in Europe. 1689 Anthony Nelme opened his workshop at 9 Ave Maria Lane; 1722–39 Francis Nelme succeeded him; then 1739–56 Thomas Whipham, formerly of Foster Lane; 1756–75 the partners Thomas Whipham jr and Charles Wright; 1775–86 Charles Wright, Whipham having retired to retail shop business. About 1786 lease of 9 Ave Maria Lane assigned to Henry Chawner, whose father, Thomas Chawner, was master to Edward Barnard at Paternoster Row and Amen Corner. 1786–96 Henry Chawner in charge with Edward Barnard his foreman; 1796–8 Henry Chawner and John Emes; 1798–1808 Emes; 1808–28 Emes's widow and Edward Barnard, with Henry Chawner jr – most women 'silversmiths' inherited an interest in their firms in this way, and often did no more than book-keeping; 1829–51 Edward Barnard and Sons (Edward John and William), removed to Angel Street 1838; to Fetter Lane 1898. 1910 became limited company, Edward Barnard and Sons Ltd, managed by Walter Michael and Stanley Barnard with George W. Joynes. 1919 moved to present address 54 Hatton Garden. Directors now Eric Barnard and G. F. Joynes. Firm specializes in heavy traditional silver which it supplies to many of London's smartest retail shops, and in elaborate cast models such as are often popular for regimental plate. Plate 319. Page 137.

Baxendale, R. G. 1912–
Britain, Birmingham

Silversmith. Influential teacher at Vittoria Street School for Jewelers and Silversmiths. Plates 366–7

Beaumont: L. Beaumont and R. Finet
France

17 rue de la République, Lyons. Retail and manufacturing jewelers and silversmiths. Founded 1830 on banks of the Saône. Moved 1894 to present address in main shopping street of France's second city. Five generations of family have managed firm, present head being Jacques Beaumont. Workshop, for half a dozen jewelry craftsmen, adjoins retail shop and produces customers' special requirements, but most of stock, as is usual with provincial retailers, is bought from the big trade factories. 1966 British Week Lyons celebrated with small exhibition from the Worshipful Company of Goldsmiths. Beaumont regularly participate in fashion shows and exhibitions staged by the three-year-old group of leading Lyons shops called *Commerce et Qualité*. Page 73.

Becker, Friedrich 1922–
Germany

Own workshop in 1952 in Düsseldorf, after three years' study at art school there, where he is now head of metalwork. One-man show Goldsmiths' Hall London with important illustrated catalogue 1966. For his fine rings, see *Goldschmiedezeitung* no 3/4 1967. Designs for Pott of Solingen and others. Plates 349, 354–5, Page 76.

Behrens, Peter 1868–1940 Germany

Architect and most influential designer, studying first at Karlsruhe then Düsseldorf. Founder member of Vereinigte Werkstätte. 1900 invited to Darmstadt, 1903–7 director Düsseldorf art school, under him perhaps the best in Germany. 1905 abandoned Jugendstil for functionalism. 1907–12 chief architect and design adviser to the big electrical firm AEG, for whom he did products, advertising and factories including their 1909 turbine factory 'the first modern building'. Gropius his pupil. 1922–36 taught Vienna, 1936 Berlin. Plates 282–3 286–8 291–2 page 134.

Benney, Adrian Gerald Sallis 1930– Britain

Artist-craftsman, known as Gerald Benney (see pages 58 98–9 and Viners). Plates 107–8 128 188–236 422 end paper

Benson, J. W. Limited Britain

25 Old Bond Street, London. Retail jewelers founded 1874, merged about 1897 with Alfred Benson and Henry Webb, who in 1889 had acquired Hunt and Roskell. Hunt and Roskell in 1846 had absorbed Storr and Mortimer, and were with Rundell Bridge and Rundell probably world's first great retail firm in modern sense, standing between manufacturer and public. Supplied much plate to royal family and other important mid-nineteenth-century British patrons; now continue as a subsidiary of J. W. Benson, themselves part of the British Silverware group.

Benson, W. A. S. 1854–1924 Britain

Architect, designer of silver and lamps, friend of William Morris. Set up factory in St Peter's Square, Hammersmith, in 1880, opened showrooms partly designed by him at 82–3 New Bond Street, though continued to show at Morris & Co of which he became a Director after Morris's death in 1896, and for which he regularly designed furniture. Helped found Art Workers' Guild 1883 and Arts and Crafts Exhibition Society 1888 with whom he often exhibited. Registered marks at Goldsmiths' Hall 1898 and 1901, some years after starting to manufacture silver – like several other arts and crafts enthusiasts disliked the hall-mark because it might damage his pieces. But favourite metals were always copper and brass. Pioneer of machine production, his trim work, often using standard parts, was generally admired: Bing sold it in 'la Maison de l'art nouveau', Paris, including it in his Paris 1900 World Fair display; it was shown in Hirschwald Gallery in Berlin; Trondheim Museum bought it liberally; and *Magazine of Art* 1896 called it 'palpitatingly modern'. 1893 published *Elements of Handicraft and Design*. Designed pavilion at the Glasgow 1901 exhibition. Wound up firm in 1920 after war munitions work. 1925 *Drawing, its History and Uses* published posthumously. (See Shirley Bury in *Country Life*, 18 March 1965 see page 67.)

Beran, Gustav 1912– Holland

Born Vienna where studied under Joseph Hoffmann and Eugen Mayer. Now art director and head designer of Van Kempen and Begeer.

Bernadotte, Sigvard 1907–

See Jensen, pages 11 137, plates 14 15 18

Bindesbøll, Thorvald 1846–1908 Denmark

Entered Royal Academy in 1861; architect 1876. Designer of textiles, furniture, book bindings, metalworks, etc. Was in his time most prominent ornamental artist in Denmark working independently of *art nouveau* movements in neighbouring countries. Designed logotype trade mark and labels still in use for Carlsberg beer. Plates 310–13, page 10

Black, Starr & Frost USA

Fifth Avenue, New York. Founded 1810 New York by Isaac Marquand on Lower Broadway; 1876 moved to Fifth Avenue; moved to last premises at 594 Fifth Avenue 1912. Named successively: Marquand & Co, 1810; Ball, Tompkins & Black, 1839; Ball, Black & Co, 1851; Black, Starr & Frost, 1876; Black, Starr, Frost-Gorham Inc, 1929; Black, Starr & Gorham, 1940. 1810 New York's population about 175,000 and only two streets were paved; new kerosene lights were fashionable and firm featured them prominently in their advertisements with porcelain and silver; firm's shop at 247 Broadway, surmounted by a gilt eagle, was known as 'sign of the golden eagle'. 1851 big crowds saw the nineteen gold-sheathed swords made by Ball, Black & Co. for presentation by State of Illinois to officers who had served in the Mexican War; in same year a tea service 'made of pure gold but six months extracted from the mines of California' was given by group of New York merchants to Edward K. Collins who had established a trans-Atlantic steamer line; 1858 firm celebrated laying of trans-Atlantic cable – in which it had itself invested – selling sections of the cable and displaying a drawing over its entrance of two huge hands clasped in greeting, symbolizing England and America. 1860, Lloyd's *Pocket Companion Guide* wrote: 'Constructed . . . of East Chester Marble, it presents an ornament at once striking and beautiful, and may well be called "The Diamond Palace of Broadway".' It had a new steam-driven elevator called a "dummer". The Prince of Wales was reported in the same year: 'The Prince has been here; has danced, supped, been fêted – bought pearls at Ball & Black's, and been to Barnum's Museum – done New York, in fact.' 1861 firm advertised in *New York Daily Tribune* offering 'camp chest and Ball's American Camp Cooking Range and Boiler to officers of Army and Navy . . . Cooking Range and Boiler made of the best Russian iron and very durable . . . indispensable articles . . . conducive to their health and comfort in the camp'. Sold to many famous millionairesses such as Vanderbilts, Guggenheims and Carnegies. But during present century the firm has specialized in precious goods. Christmas 1959 made deliveries by helicopter to its suburban stores at White Plains, Manhasset and Millburn. 1960, the oldest surviving retail firm in New York except for a small Fulton drug store, celebrated 150th anniversary by displaying treasures of Worshipful Company of Goldsmiths, London. Diamond-International Awards 1960 and 1962. Now owned by Marcus Purchasing Company. Fifth Avenue business closed 1965 will reopen soon at new address.

Blyth, Ernest A. 1939–
Britain

Goldsmith and jeweler. Finished apprenticeship in Goldsmiths' Hall Assay Office 1961. Part-time student at Central School of Arts and Crafts from 1957. Designed range of silver jewels for George Tarratt Ltd, Leicester 1962. Worked for some time with H. J. Company Ltd. Now has own workshop in White Horse Street, London, teaches Hornsey College of Art.

Bojesen, Kay 1886–1958 Denmark
Plates 150 340 342

Bolin, W. A. Sweden

Stockholm court jewelers. See pages 9 23–25 73 142, plates 32–42 124–25 and front cover

Bonebakker & Son (As. Bonebakker & Zoon)
Holland

Crown Jewelers. Founded by Adrian Bonebakker, b 1767 Tiel. Probably studied in nearby Schoonhoven, then as now chief Dutch centre of silversmithing. 1792 arrived in Amsterdam to seek fortune. There were then as members of the guild only thirty-nine goldsmiths, ninety-four jewelers and 126 silversmiths – 259 artisans for population of 200,000. Bonebakker's accounts, preserved since 1805, record notable orders: presentation sword given by his friends to Baron David H. Chassé, who had fought at Waterloo and commanded Dutch troops in the citadel in 1830 siege of Antwerp. 1837 Bonebakker moved to 'The White Horse' on the fashionable Heerengracht canal. 1840 made crown for King William II of Orange. Sixth generation of family now manage firm, namely Ferdinand, the son of Adrian, nephew of Charles. 1954 moved to present shop on Rokin. 1967 celebrated 175th anniversary of foundation. Business mostly retail with only small jewelry factory.

Bonvallet, see Cardeilhac

Boucheron
France

Place Vendôme, Paris; London; Biarritz; New York. Court jewelers, founded by Frederic Boucheron at Palais Royal in Paris, later became internationally known in Place Vendôme, in New Bond Street, London and in Biarritz. Since 1858, three generations of Boucherons have succeeded each other and continued to conduct business personally. Sixty-nine years ago opened their London house.

Bradbury, Thomas & Sons Britain

Leading Sheffield firm founded c. 1777 absorbed Atkin c. 1947.

British Silverware Limited
Britain

The biggest British manufacturing and retail group, built up since 1963, includes Mappin & Webb, Elkington, Adie Bros, Walker & Hall, Garrard, Benson, and Manoagh Rhodes in Bradford. Owned jointly by Mr Charles Clore's Sears Holdings and Delta Metal. Chief designer Eric Clements.

Broggi (Fabbrica Argenteria Broggi)
Italy, Milan

Founded 1842 by Gaetano Broggi, small artist-craftsman who did work at Court of Savoy for King Carlo Alberto. Joined by his two sons Carlo and Serafino and through their efforts firm grew rapidly. 1866 moved and began to manufacture spoons and forks with modern machinery. Since then firm has moved several times, now produce wide range of tableware and have factories in several parts of Italy.

Bruckmann, P. & Söhne
Germany, Heilbronn

Founded 1805 by Georg Peter Bruckmann – who was first apprenticed to his father, later studied at Academy of Vienna, Académie Julian, Paris, and in Geneva. 1820 erected first major press to increase output and reduce price of silverware. At his death in 1850 his son Ernst Dietrich Bruckmann (1829–70) returned from America and took over management. 1864 introduced modern plant with steam engine to increase production of spoons and forks and by 1885 annual turnover more than 1·5 million marks. During this expansion a school was founded to train artist-craftsmen, which grew into Professional Bruckmann School for Designers, Chasers and Engravers and Silversmiths. Towards end of 19th century new designs introduced, getting away from 'reproduction' designs, expressing the 'Jugendstil' later influencing the 'Deutscher Werkbund'. Plant rebuilt 1944 under Dietrich Bruckmann – fourth generation. Many prizes. Plates 283 284

Buccellati, Mario

Italy

Via S. Margherita 5, Milan; Rome, Florence, New York, Palm Beach. Court jeweler. Mario Buccellati (b 1891) studied goldsmithing from 1903 at Beltramie Besnati. 1919 acquired and carried on business under own name. Supplied Popes Pius XI and XII, the royal families of Italy, Spain, and Egypt, Italian Government and Gabriele d'Annunzio who from 1921 to 1937 wrote hundreds of letters to his friend and supplier Mario, calling him in one of them 'Mastro Paragon Coppella' – master of precious metals – which phrase is now the firm's trade mark. Specializes in, and possibly invented, the textured engraved finishes worked with the *bulino* or chisel, for which Italy is now famous. Employs over 200 craftsmen. Silver now sold in some of the more exclusive US department stores like Neiman-Marcus of Dallas, Texas. See pages 53 136

Bulgari

Italy

Retail and manufacturing jewelers and silversmiths. 1850 George Boulgaris opened small silver factory on Turco-Albanian frontier frequented by the beys of Albania; after 1877 massacre of Christians, escaped to Corfu with his 19-year-old son Sotirio (b 1857), the only survivor of his nine children. Soon after Sotirio and friend went to Naples and opened tiny shop in Piazza dei Martiri selling elaborate silver, daggers, etc., inspired by Turkish and Iranian work then popular. Store was robbed and the two friends went penniless to Rome. 1881 started selling trinkets on stand outside the Frency Academy. Friends from Greek colony in Rome helped by selling Bulgari craftsmanship in their stores. Sotirio opened shop first in Via Sistina, then at present site in Via Condotti – but still got up at dawn to melt silver for his day's work. Returned to Greece, married, and in 1905 opened a real jewel showroom. His sons, Constantino and Giorgio, had taken on management by the time he died in 1932. Giorgio increasingly interested in writing: his fine series of books *Argentiri, gemmari, e orafi d'Italia* are now appearing. 1934 shop was remodelled to its present shape. Today Anna and Marina, Constantino's daughters, supervise workshops and shop; Giorgio's sons Gianni, Paolo and Nicola direct buying and selling. Bulgari are world famous for their fine stones and the richness of their collections both antique and modern. See page 136

Burch-Korrodi, Meinrad 1897–

Switzerland

Artist-craftsman with retail shop, Zürich. See pages 93 95 141, plates 157–66

Calderoni

Italy

Via Monte Napoleone, Milan. Court jewelers. Founded 1840 by Adone Calderoni. Up to 1910 production and retail sale in centre of Milan. 1911, the first firm in Italy, started to sell its production throughout country. 1925 enlarged jewel factory to employ 30 workmen, became public company. 1927–8 shops in Bari, Messina, Livorno, Turin, Asmara, Addis Ababa. Exhibited frequently. Organized 1958 and 1959 international exhibitions of jewelry and silverware. Diamonds-International Award 1960. See page 136

J. E. Caldwell & Co.

USA

Founded by James E. Caldwell (1805–81). Watchmaker and jeweler, who left Poughkeepsie, New York, to start his own business in 1835 or 1839 on Chestnut Street, Philadelphia. 1843–8 James M. Bennett partner. One price system used from 1840s, unusually early for USA. Several changes of premises, always on Chestnut Street. Pioneers of advertising in *Godey Ladies Book*, *Harpers Weekly*, *Saturday Evening Post*, *McClures* etc., and of staff welfare resulting in early closing days. 1869 the six young night guards woken possibly by burglars trying to blow the safes; huge fire in which two guards died. Subsequent press claim 'Our "fire-proofs" saved us $125,000 in diamonds and watches besides nearly $400,000 in bonds, notes, deeds, and other valuable papers, many of which belonged principally to our customers.' In 1876 there were only some 5,000 'Bell Speaking Telephones' in USA and Caldwells were one of the first businesses in Philadelphia to use the new wonder. Showed in the 1876 Centennial exhibition. On James's death in 1881 his son J. Albert succeeded. Record sales in 1865 because of Civil war, and in 1898 because of Spanish American War. Many silver commissions for US navy. 1905 and thereafter, made hero medals for Andrew Carnegie's new Hero Fund Commission. The store's eighteen horses named after stones like Sapphire, and fourteen wagons, were all replaced by vans before 1914. 1914 on Albert's death, J. Emott Caldwell succeeded till his death 1919. Moved to present address Chestnut and Juniper Streets 1916. Rich regency style interior, architect Horace Trumbauer, designer of Philadelphia Free Library, William R. Eisenhower new manager, then Charles W. Oakford. In slump, 1932 sales sank to record low level of 1893 or 1850s. Much short time in factories, sometimes no jewelers there at all. Stationery made in Belmont Building, North 13th Street. 'The Book' kept in the store is socially important: some two hundred debutantes' names are listed, with the dates of Charity Ball, June Ball, First City Troop Gala, carrying on pre-independence traditions. Austin Homer President 1952; 1953 new store in du Pont Hotel, Wilmington, Delaware. 1954 new store at Haverford, both successful and since enlarged. 1963 opened Steuben glassroom. 1964 lamp department discontinued. Now one of Philadelphia's biggest jewelry factories.

Cardeilhac
France

8 rue Royale, Paris. Founded 1802 by Vital Antoine Cardeilhac. 1817 started using his hall-mark – the Legion of Honour cross. 1819 the firm moved to large premises and by 1823 became known as important traders. Also started to manufacture heavy solid silver articles. 1850 his son Edward took up the business. 1860 Ernest Cardeilhac became head of firm and exhibited with success in world exhibitions at end of century, producing sober type of saleable *art nouveau*. 1889–1931 produced items which have become representative of period. Bonvallet was his best designer and much of their work can be found at the Musée des Arts Décoratifs. 1904 Cardeilhac died, leaving widow and two young sons to continue. From 1927 James and Peter went from success to success, winning many awards. 1944 Peter died, James became sole manager. 1951 amalgamated with Christofle, James Cardeilhac and principal members of staff continuing in the firm. Plate 298

Carlman, Sven 1901–
Sweden

Owns and directs old family firm of jewelers in Stockholm; accomplished musician. Designed flatware which is hand-made by his firm C. F. Carlman AB. With Bolin, is purveyor to Royal Swedish Court.

Cartier
France

13 rue de la Paix, Paris; New York; Palm Beach; Caracas; London; Cannes; Monte Carlo. Court jewelers. About 1897 in Paris, Louis Francois Cartier started one-man jewelry workshop in a garret. Creations attracted attention of Her Imperial Highness, Princess Mathilde Bonaparte, and with her invaluable patronage Cartier was able to establish himself in 1859 in modest retail shop on the Boulevard des Italiens. 1874 joined by son Alfred, whose own sons – Louis, Pierre and Jacques – moved shop to 13 rue de la Paix at turn of century. The Prince of Wales, during one of his visits to Cartier, Paris *c.* 1900 remarked that it was unfortunate not having a branch in London. 1902 branch was established in New Burlington Street. After Great War it was formed into an English limited company and in 1921 Cartier Ltd set up manufacturing workshops in London: English Art Works Ltd for jewels, and Wright and Davies Ltd for gold small work. 1912 Pierre founded New York house, sold 1963. Page 73

Chaumet
France

Place Vendôme, Paris; New Bond Street, London. Founded 1780 by Etienne Nitot, later Napoleon's jewelers, making crown given by him to Pope Pius VII, his sword, etc. 1875 M. Joseph Chaumet opened London house. Continuous expansion – major awards at Paris 1900, St Petersburg, St Louis 1904, Milan 1906, Buenos Aires 1910, Paris 1925, 1937, and Brussels 1958. 1928, on M. Joseph Chaumet's death, M. Marcel Chaumet succeeded to control of firm. He and his family are now the principals. The firm specialized in tiaras and in jeweled ceremonial clothing for the orient. 1967 agency agreement with Wolfers of Brussels.

L'Orfèvrerie Christofle
France, Paris

Founded 1839 Paris by Charles Christofle. Founder of French plated and gilded silverware industry, soon manufactured own gold and silverware. Spent whole personal fortune of half a million francs buying all French electroplate patents (see Elkington), and raised a further 1½ million francs from friends. Made a fortune, was awarded Cross of the Legion of Honour, and by 1859 employed 1500 people. Won highest awards at Paris Exhibitions in 1839, 1844, 1849. 1867 exhibition showed works commissioned by Emperor Napoleon III, Empress Eugenie, Imperial Court and City of Paris. 1855 Pauline de Metternich, wife of Viennese Ambassador in Paris, dined with the emperor. She complimented him on beautiful silver and he said: *Chère ambassadrice, c'est là un luxe que tout le monde peut désormais s'offrir. Je n'ai en effet sur ma table aucune pièce d'argent massif. Elles sont toutes en Christofle.* Electroplate had seduced even the head of state.

Due to activities of his nephew, Henri Bouilhet, pieces were also shown abroad: London 1851 and 1862; Vienna 1853; Philadelphia 1876. Christofle was one of first silversmiths to use stamped mark on goods as guarantee of weight of silver used. Present Director and President of company is Tony Bouilhet, grandson of founder's nephew. Factories established in Switzerland, Buenos Aires 1946, Milan 1955, New York 1958.

'Baguette' is probably the commonest flatware pattern in Europe, and the most satisfying to use. Christofle first produced it in 1861, and are still selling it in 1967 at 100,000 pieces annually. Plates 438 441, page 142

Citroën
Holland

Amsterdam Kalverstraat 1. Court jewelers founded 1827. Now managed by Karel Citroën (b 1908), sixth generation of family, *art nouveau* collector and dealer, much of whose first collection, after exhibition throughout Europe, was sold to Darmstadt Museum. Lecturer at Amsterdam University. For British Week Amsterdam 1966 showed modern jewels by Andrew Grima of London. Associate of Worshipful Company of Goldsmiths, reader in art history, Amsterdam university. The founder was Barend Roelof Citroën, goldsmith; the firm moved to its present site in 1859. 1927 appointed jewelers to Her Majesty The Queen of the Netherlands.

Clarke, Geoffrey 1924–
Britain

Sculptor, stained glass artist, printmaker. Studied Lancaster College of Art, Royal College of Art, teaches Chelmsford; country house studio, workshop at Stow Hill, Stowmarket, Bury St Edmunds. Largely responsible for recent revival of British stained glass because of impact of his windows at Coventry Cathedral. Windows also in Worshipful Company of Goldsmiths' small treasury at Lincoln Cathedral, Plymouth Crown Hill, Taunton Crematorium. Large commissioned sculpture, usually in cast aluminium, at Castrol House and Thorn House, London; Principal's Lodging Newnham College Cambridge; and at Durham, Newcastle, Minneapolis St Paul, etc. Church metalwork at Coventry Cathedral, King's Lynn St Margaret's, Chichester Cathedral, Shere (with Louis Osman), Guards Chapel London, Winchester College, St Louis priory, Missouri. Set of silver tankards given by Worshipful Company of Goldsmiths to Lancaster University 1967. Plates 414 416–18 title page and frontispiece, page 131

Clements, Eric 1925–
Britain, London

Artist-craftsman. Studied Birmingham College of Art and Royal College of Art 1949–53. Won Worshipful Company of Goldsmiths' national competitions Coronation plate for Southend-on-Sea 1954; 1956 for a jewel for Mayor of Bolton. Ford Foundation ESU Fellowship USA 1958. Specializes in ceremonial silver designs such as coffee set commissioned by HM the Queen for King and Queen of Denmark 1957, and sword of office commissioned by Company for Chief of Air Staff, RAF 1959. Consultant designer Mappin & Webb Ltd and other industrial concerns. Head of Industrial Design School, College of Art, Birmingham. Plates 373–7, 379

Clen-Murphy, Desmond 1924–
Britain

Artist-craftsman. Studied Brighton College of Art 1946–51; assistant to Gerald Benney, London; taught Brighton, Worthing, Chichester; own workshop Brighton since 1959. Plates 402–3 411–13

Cohr, Carl M.
Denmark, Fredericia

Manufacturing and retail silversmiths. Founded 1860 by Ditlev Cohr. 1863 son Carl M. Cohr took over shop and workshop consisting of only two journeymen. Success followed extension of sales to neighbouring Kolding, Vejle and Middelfart. By 1900 35 craftsmen, by 1907 there were over 70, and by 1918 several hundred. 1895 silver holloware added to original range of flatware and engagement rings. 1921 silver plate introduced with trade name 'Atla', followed in 1930 by stainless steel. Many exhibitions. Designs produced by Knud

von Engelhardt (architect c. 1908 of Copenhagen trams) are in Kunstindustrimuseum Copenhagen. Other artists employed include sculptor Sigfred Wagner, H. P. Jacobsen and Hans Bunde. By 1967 Cohr had produced 88 different cutlery patterns. Now uses 90 tons annually of raw material – gold, silver, nickel and steel. Exports overseas started 1921. The 'Old Danish' pattern of cutlery introduced 1860 with threaded edge, now simplified, is still one of firm's best sellers. Third and fourth generation of original family now manage firm with Einar Cohr as head. Since 1860 some 43 million pieces of flatware have been produced. The world's appetite for flatware is incomprehensible. It is an unbreakable, very long-lasting and rather unbeautiful commodity; yet Cohr, for example, one of several Danish factories of only modest size, have made five times as many pieces as there are people in Denmark. Plate 309, pages 11 142

Comyns, William & Sons
Britain, London

Manufacturers, producing fine hand-made silver over long period. Founded 1848 by William Comyns in Silver Street, soon moved to Beak Street in building which had once been Canaletto's studio. Sons Charles (d 1925) and Richard (d 1953) continued, and in 1953 Bernard Copping acquired firm with 30 people, moved to Dean Street, then to Tower Street. 1967 Comyns employs 60 people, weekly turnover often equals a whole year's in 1950. Bought Charles Fox patterns 1920. See page 74

Cooper, Francis 1906–
Britain

Learnt craft in father's workshop at Betsoms Hill, Westerham. Notable examples of work are mustard pot for Corpus Christi College, Oxford, and mace for Farmers' Company.

Cooper, John Paul 1869–1933
Britain

Architect, silversmith and jeweler. Educated Bradfield College; 1887 articled first to J. D. Sedding, then to Henry Wilson. On Wilson's advice took up metalwork 1897, reviving use of shagreen. Taught at Birmingham 1904–7, later setting up workshop in Kent. Succeeded by son Francis Cooper.

Craver, Margret (Mrs M. C. Withers, b Pratt)
USA

Middle Road, West Newbury, Mass. Artist-craftsman. Works owned by Newark Art Museum, Museum of Contemporary Crafts. Consultant to Occupational Therapy Dept US Army during 1939–45 war, and head of Hospital Service Dept, Handy & Harman. Invited to Worshipful Company of Goldsmiths' Conference, London. Headed Handy & Harman's Education Dept which conducted conferences for American art teachers. Designed exhibition 'Handwrought Silver' which opened at Metropolitan Museum of Art. It was later purchased by US State Dept and circulated in USA and overseas. Wrote technical booklets for silversmiths and jewelers. Co-authored and did research for cover article 1964 August magazine of International Institute for Conservation of Artistic and Historic Works.

Cuzner, Bernard 1877–1956 Britain

Artist-craftsman. Educated Redditch School of Art; then Vittoria Street School for Jewelers and Silversmiths under R. Catterson-Smith and Arthur Gaskin 1896–1900. Afterwards taught there. Retrospective exhibition Malvern, England, 1960. His workshop is basis of an exhibit at Birmingham Museum of Industry. Page 137

Czeschka, Økart Otto 1878–1960
Austria

Artist-craftsman, graphic designer, painter. Studied at Vienna Academy, taught at Kunstgewerbeschule. Member of Wiener Werkstätte. A fine candelabrum by him c. 1920 is in the Hamburg Museum für Kunst und Gewerbe. Plate 261

Dentler, Rudolf 1924–
Germany

Studio in Blaubeuren. Trained at art schools in Germany and Switzerland. Since 1955 has been independent designer-craftsman at Blaubeuren; his first exhibition at the Ulm Museum in 1958 led to showings throughout Germany, France, England and Switzerland. Designed jewelry from 1960, shown Paris 1961 and at Goldsmiths' Hall London international exhibition. Now uses scrap such as watch parts and broken trinkets to obtain decorative surfaces.

Devlin, Stuart 1931–
Britain

Born in Australia. Worked for two years as designer-craftsman in silver at T. Gaunt & Co., Melbourne; taught in art schools in Victoria; awarded Diploma of Art from Royal Melbourne Technical College; grant in 1958 to study in Britain. Studied at Royal College of Art in London; awarded unprecedented double DesRCA and thesis prize. Work in silver exhibited by Worshipful Company of Goldsmiths in many countries. Awarded a Harkness Fellowship of the Commonwealth Fund of New York in 1960 for two years' study and work in the USA. Exhibited in New York at Thibaut Gallery 1961. Also designed the Australian decimal coinage. One-man show Terry Clune Gallery Sydney, director of art education Victoria. Now resident permanently in London, where he has designed and made many important commissions such as maces for Universities of Melbourne and of the Pacific at Los Angeles and of

Bath in England; still designs medals, notably a series with Benney of six British prime ministers' heads for Medallioners Ltd. Plates 398–401 428–30

Dittert, Professor Karl 1915–
Germany

Silversmith and designer in Schwäbisch Gmünd, studied in Berlin and Schwäbisch Gmünd under Professor Warneke. Industrial design, in particular for WMF. Third prize international competition 'silver coffee pot' 1960. Judge for De Beers Diamonds-International Awards meeting Goldsmiths' Hall London 1967. Plates 83–6 102 140

James Dixon & Sons Limited
Britain, Sheffield

Silversmiths. Founded 1806 by James Dixon, the firm was among first in Britain to manufacture Britannia Metal and Old Sheffield Plate, as well as solid silver. 1822 firm moved to larger premises at Cornish Place, Sheffield. Name refers to their pioneering and large scale use of Cornish tin. Installed most up-to-date machinery. Although firm has grown and grown, and more buildings have been added to original factory, it is still situated at Cornish Place. Firm became known by present title after Dixon was joined in business by sons William Frederick, James Willis, and Henry Isaac, and by son-in-law, W. Fawcett. James Willis was responsible for American side of the business and for a time lived in New York. 1930 Dixon's took over goodwill of William Hutton & Sons. Firm now has showrooms in London, Australia, Italy, New Zealand and Switzerland, and is run by fifth generation descended from founder. See page 75

Donald, John 1928–
Britain

Studied graphic art Farnham School of Art, and jewelry Royal College of Art. Won competition for design of Worshipful Company of Goldsmiths' Wardens' badges and made them. 1961 own London workshop. 1963 second prize, German Goldsmiths' Society 30th anniversary international competition. Took exhibitions of Worshipful Company to Hudson's Bay Store, Vancouver 1966, Jordan Marsh Store Miami Florida 1967. Teaches Royal College of Art. 1967 opened new shop designed by Alan Irvine and workshop in Cheapside where Goldsmiths' Row once was, near Goldsmiths' Hall in old City of London, the region of craft workshops in Middle Ages. See 'Donald' by Graham Hughes, article in *Connoisseur* October 1963. Plates 129 410

Dragsted, A.
Denmark

Bredgade 17, Copenhagen. Jewelers and silversmiths to Royal Court of Denmark and Greece, and to Imperial

Court of Russia. Established 1854 by great-grandfather of present principal, whose cousins carry on a silver factory under name E. Dragsted. Freelance designers used, but silver supervised by Folmer Dalum and jewelry by Jørgen Klint. Many exhibitions. See Strand. One of few firms in Scandinavia who sell precious as opposed to silver jewelry. Plate 307. Pages 11 142

Dresser, Christopher 1934–1904, Britain

Perhaps the most original British designer of silver in 19th century. Studied from the age of 13 to 15 at the School of Design at Somerset House, then as a botanist. 1860 became botany lecturer at the Department of Science and Art and at St Mary's Hospital Schools, having published articles on plant forms and structure 1857–8. Began designing silver early in 1860s, working mainly for Hukin and Heath of Birmingham and James Dixon of Sheffield, also occasionally for Elkington. In late 1870s he organized for Hukin and Heath, electrotype copies of Japanese and Persian metalwork; also opened an oriental warehouse in Farringdon Road with Charles Holme of Bradford, and in 1880s designed and exhibited furniture at the Art Furniture Alliance in Bond Street. Visited Japan in 1876. Developed an extremely logical and functional style, explained in a series of articles on the principles of design in *The Technical Educator* 1871–2. He was important not only because of his art; but also because, unlike most theorists of his time, he had the courage to practise it successfully in industry. Plate 237

Durbin, Leslie G. MVO LLD 1913–. Britain

Student at Central School, London, and carried on in his own workshop at Rochester Place, Camden Town. Taught at Central School of Arts and Crafts and Royal College of Art, London. Made many masterpieces for City Companies, the Worshipful Company of Grocers centrepiece and coat of arms, Guildford Cathedral Chapel of Chivalry plate, badges of office and maces. Designed plate for Hudson's Bay Company, Canada; altar set for Southport Church, Connecticut; president's badges for American Orthopaedic Association; the Association of Orthopaedic Surgeons; the New York Microscopical Society; and to celebrate its 1965 centenary, the mace and president's badge for the Smithsonian Institution, Washington DC; 1967 mace for the University of South Carolina. Own workshop London. Plates 112–13 368 370–2, page 137

Eisenloeffel, Jan 1876–1957, Holland

Versatile designer Amsterdam, worked Munich 1908.

Britain, Birmingham Elkington

George Richards Elkington, 1801–65, apprenticed at 14 to his two uncles, Josiah and George Richards, to learn silver plating. This meant 'close plating' – hammering thin layers of silver over base metal such as copper, adhesion being by soldering and rolling. Soon became a partner, becoming sole proprietor when uncles died. Eventually took cousin Henry Elkington into partnership, introducing new processes and techniques, upon which John Wright, surgeon inventor, and Alexander Parkes, metallurgist, also worked. Between 1836 and 1838 they took out various patents, including one for 'electro-gilding', and are, therefore, rightly considered inventors of electroplate. Built up the business on scientific basis, despite opposition from established trade. 1841 a large works was completed in Newhall Street, Birmingham, which for some years after 1945 served as Birmingham Museum of Industry, till its demolition in 1966. Fresh capital being essential they admitted Josiah Mason as a partner; he had prospered as a steel pen manufacturer, was knighted in 1873, and subsequently benefited from his association with electroplating. Gradually electroplated articles became popular all over the country, and graced almost every Victorian table. Firm's products displayed at the Great Exhibition of 1851, resulting in even more commissions at Birmingham works. 1860 invention of dynamo revolutionized process of electroplating. The French electroplating pioneer, de Ruolz, challenged Elkington's 1842 French patent application; settlement was reached and honours of invention shared. The French industry was supported by M. Christofle, and attained great economic importance.

Henry Elkington died in 1852, having proved his worth particularly on artistic side of the business. Original founder carried on for a further 13 years, setting up a great copper refining enterprise at Pembrey, Carmarthenshire. Business passed to his son when he died in 1865.

Silver plating was invented in 1840. Electric current would then have been obtained only from batteries, but already in 1845 the magneto electric machine of J. S. Woolrich, using principles discovered by Faraday, was in use (Huttons of Sheffield had a Woolrich licence as late as 1945) and this was followed about 1866 by Wilde's dynamo machine: scientific development in which Elkingtons were continuously the world pioneers. For instance Sir William Siemens in his biography pays tribute to the important part played by Elkingtons in starting both Siemens and Krupp in these industries. See Wiskemann, Christofle, Hutton and Walker & Hall; see R. E. Leader, Institute of Metals journal number two, 1919, and R. S. Hutton, Hothersall Memorial Lecture 1959, transactions of the Institute of Metal Finishing Vol 36.

January 1963 Elkington's and Mappin & Webb merged and formed British Silverware Ltd. Shortly afterwards the new company acquired Walker & Hall, Adie Bros., and Gladwyn Ltd. Products of all four firms are today sold under the Elkington name. The machines from the Walker & Hall Bolsover factory have been moved to Walsall, Bolsover made into a warehouse, and the central Sheffield Walker & Hall factory demolished. In 1967 Elkington's factory at Goscote, near Walsall, closed, and with it the firm's history as a separate unit, with its 250 men.

Emerson, A. R. MBE 1906–
Britain

First apprenticed in the trade while working at Central School; later became instructor in silversmithing and eventually Head of the Department. Did valuable work during war in starting a machine shop at the School for production (which led to a new department) and for International Apprentices' Competition. Now teaches at Brighton, head of metal-work at Sir John Cass College, London's chief training centre for craftsmen.

Ercuis
France

Founded 1867 under name of Pantographie Voltaique; firm now takes its name from its location – Ercuis, a small town in the Oise department. For 30 years firm devoted itself exclusively to producing ecclesiastical articles then, at the turn of the century, began to manufacture silver tableware, now its principal trade.

Eriksen, Sigurd Alf 1899–
Norway

Apprenticeship at Tostrup, Oslo. Trained for enamel work at Simet & Co., Vienna, at Ulbrich, Hanau, at Schneider, Paris. Trained in enamel miniature painting at Heinrich Geier, Paris. Attended State School of Arts and Crafts (Statens Handverks-og Kunstindustriskole), Oslo, and State Academy of Drawing, Hanau a/M. Studied painting at Academy at Hanau, and at school of Professor Ewald in the same town. Travelled and exhibited extensively. Prix d'honneur at Milan Triennale 1954; Tostrup medal for Applied Art 1957. Awarded Norwegian Association of Designers' annual award 'Jacob' 1961. Taught at schools in USA and in Copenhagen. Now teacher at State School of Arts and Crafts, Oslo, and runs own workshop.

Fabergé
Russia

Goldsmiths and jewelers. Established St Petersburg 1842 by Gustav Fabergé (1814–93). His son Peter Carl Fabergé (1846–1920) took control in 1870 and in 1882 was joined by younger brother Agathon. 1884 firm received Royal Warrant of the Tsar Alexander III. This was the year when first of famous Imperial Easter eggs was made. Achieved great success through their high standards of craftsmanship. 1882 gold medal, Pan Russian exhibition Moscow. 1885 gold medal Nuremberg. 1888 special award Copenhagen. 1900 Carl Fabergé awarded Légion d'Honneur after Paris exhibition. Many other awards. Branches established in Moscow (1887–1918), Odessa (1890–1918), Kiev (1905–10), London (1903–15). 1910 unsuccessful test case against Worshipful Company of Goldsmiths, disputing British hall-marking law. Closed by Bolsheviks 1918. Carl Fabergé died Lausanne. At the height of their fame, staff of over 500, including some 30 designers, but Carl and Agathon Fabergé exercised close personal supervision and designed the majority of pieces made by firm in St Petersburg workshops. See Snowman, *The Art of Carl Fabergé* and pages 23 47 49 136, plates 89–92

Fenster, Fred 1934–
USA

Educated City College of New York and Cranbrook Academy of Art. Studio workshop in own home and now Assistant Professor of Art, University of Wisconsin. Work represented in Plattsburgh State College, NY, Detroit Institute of Art, Milwaukee Art Centre, Walker Art Centre St Paul Minneapolis. Fiber Clay Metal award 1960, Wisconsin Designer Craftsman award 1962, '63, '64, '65, '66. Represented in *Craftsman USA, 1966, American Jewelry Today, 1963*. Plate 344

Feuillâtre, Eugène 1870–1916
France

Sculptor, goldsmith, enameller. Worked for René Lalique. First showed independently with Société des Artistes Français from 1899, and at Paris 1900 and Turin 1902 exhibitions. Specialized in holloware of translucent enamel, glass and silver. See page 135

Firmin & Sons Limited
Britain

8 Cork Street, London, and Globe Works, Portland St, Birmingham. Manufacturers. Button-makers, started by Thomas Firmin (1632–97) in Three Kings Court, Lombard Street; premises rebuilt after Great Fire of London, 1666. Buttons were of silk or woven metal thread till the Royal Regiment of Artillery adopted brass in 1680. The industry had been predominantly French but under Firmins became mostly British. Firmin buttons were worn by Wellington's and Napoleon's men at Waterloo, by both sides in the American War of Independence and the Civil War and by at least eight of the allied armies in the 1939–45 war. Business carried on at the Red Lion, Strand, and in 1722 Nathaniel Firmin renewed lease as owner until his son Samuel took over in 1754. From then on detailed records were kept of purchases by many members of the royal family. Warrants are now held for HM the Queen and HM Queen Elizabeth the Queen Mother. Seems likely that these premises were on part of the site now occupied by Australia House. 1760 a move was made to the Strand, and in 1915 the business left St Martin's Lane for Cork Street. Birmingham factory opened 1882 when manufacture in more expensive districts of London began to be uneconomical. Today a small amount of high quality handwork is still done at London Head Office. Business was owned and run by members of the Firmin family in direct line of descent until it was formed into a limited

company in 1875, and since then Board of Directors has always included a majority of Firmins among its members and the family has provided a continuous succession of Managing Directors. Comparatively modern additions to the Company's activities are the manufacture of many metal pressings and stampings in Birmingham, and the merchanting of cloth and tailors' trimmings in London. As Government contractors of many years' standing, made vital war materials in two World Wars, including service clothing accessories, metal parts for bombs and wireless sets, shell fuses and radar equipment. Today they supply uniform buttons and badges to USA, Canada and most African countries. During three years following end of Second World War in 1945 nearly half the output of factory was exported and figure today is about 40 per cent of doubled output.

Fleming, Baron Erik 1894–1954
Sweden

Silversmith, head of Ateljé (Atelier) Borgila, which he founded in 1920. With Ängmann and Nilsson was the leader of his generation. Succeeded by his son Lars Fleming (b 1928) who runs firm. See pages 137 142

Fouquet, Georges 1862–1957
France

Commander of Légion d'Honneur. Classical studies; succeeded his father, Alphons Fouquet, about 1890. Georges Fouquet created new style in jewelry, known as 'Mil-neuf-cent'. Use of various coloured stones, diamonds, pearls and much enamel. Exhibited widely, including International Exhibition at Milan 1906. 1925 President of Jewelry Class in Paris Exhibition of Decorative Arts. 1937 President of Jewelry Class in Paris International Exhibition. Member of Central Union of Decorative Arts. At Christie's, London, July 1963, a masterly gold pendant without stones fetched £420: encouraging evidence that vintage quality is already attaching to such pieces of little intrinsic value, simply because of their virtuoso design. See Mucha

Fouquet, Jean 1899– France

Jewelry designer. 1919, after classical education, joined his father's firm. Showed in Paris 1925. With a number of other designers and jewelers, including Raymond Templier and Georges Bastard, founded Union des Artistes Modernes (UAM). Showed in Paris 1937. Regular exhibitor since 1926 at Salon d'Autonne, etc. Showed Brussels World Fair 1958. Author of *Bijoux et orfèvrerie* 1928, a picture review of French work. Page 135

Friend, G. T. OBE
Britain

Engraver. Teacher of engraving and lettering at the Central School; his work was regarded as supreme. His workshop was at 9 Dyers Buildings, Holborn, where Cyril Baker and others worked with him.

Fröhlich, Max 1908–
Switzerland

Silversmith and jeweler, head of metalwork at Zürich, Kunstgewerbeschule; Ring of Honour 1966 from German Goldsmiths' Society; State Prize of Bavarian Government at International Handicrafts Fair, Munich; frequent Swiss representative on international juries. Plates 356–7

GAB

See pages 9 30–31 137, plates 43–62 121

Gaillard, Lucien 1861–
France

Contemporary of René Lalique; used insect and flower forms in his silver work. Participated in the Glasgow International Exhibition of 1901.

Gardberg, Bertel 1916–
Finland

Designer. Educated goldsmiths' school, Helsinki, school of industrial art, Helsinki. Gold medals Milan Triennale 1954 and 1957; four silver medals 1960; golden ring of honour of the German Goldsmiths' Society 1960; Lunning Prize 1961. Worked Denmark 1946–9, France 1953–4. Own workshop Finland since 1949. Specializes in wood and metalwork. Exhibited widely. 1963 Galerie du Siècle, Paris. Now head of metalwork at Kilkenny Castle, the Eire government craft centre, and resident there as general art director. Plates 360–4

Garrard & Company Limited
Britain

112 Regent Street, London. Crown Jewelers, since their first appointment 1843. Descended from George Wickes, George I's goldsmith who started in Threadneedle Street in 1721 and in 1735 moved to Panton Street, Haymarket. During 19th century the Garrard family firm was one of leading British silversmiths and maintained a silver and jewel factory, as well as shop in Albemarle Street opened in 1911 for George V's coronation. There they made many of the British and foreign crown jewels. Some of very best Victorian silver, for instance, at Buckingham Palace or Goldsmiths' Hall London, was made by them. 1952 acquired by Goldsmiths' & Silversmiths' Company Ltd, who adopted the name Garrard, closed Albemarle Street factories and transferred business to 112 Regent Street. 1963 merged with Mappin & Webb, Elkington, Adie Bros. and Walker & Hall under the name British Silverware. 1963 modern silver exhibition, an important selection of 80 pieces designed over 16 years by Alex Styles, their staff designer, made under the British government scheme to assist fine work, 'Assistance to Craftsmen', started in 1947 to overcome wartime difficulties and discontinued 1963. 1966 a further impressive display of work by Styles included 40 pieces made in three years, showing that the pace of production had

increased considerably. They bought dies of Francis Higgins, London's leading hand-forging cutlers, to preserve their skill; lent them to E. Barnard and then to C. J. Vander Ltd. Plate 378

Gaskin, Arthur 1862–1928
Britain

Metalworker, painter and illustrator; played prominent part in bringing arts and crafts movement to Birmingham; specialized in enamels. Studied Birmingham School of Art then joined staff of the School. Married Georgie Evelyn Cane France, a student at the School 1894. Took up metalwork and, working with his wife, produced his first jewelry 1899. Succeeded R. Catterson-Smith as Headmaster of the Vittoria Street School for Jewelers and Silversmiths 1902. Plate 317

Gaudernack, Gustav 1865–1914
Norway

Silversmith and designer. Born Bohemia, studied Austria, came Christiania (now Oslo) 1891, designed glass for Christiania Glasmagasin, worked with the silver firm David Andersen 1892–1910 when he started his own workshop. Died of appendicitis four years later. Started with the clumsy Norwegian 'dragon' style, inspired by old Norse art, developed his vigorous *art nouveau*, finished after 1910 neo-classical. Won for Andersen two gold medals Paris 1900, Grand Prix St Louis 1904. Himself won gold medal Norwegian Centenary Exhibition 1914. Plates 300–4

Gebhart, Friedrich 1914–
Germany

Studied Worms, Berlin. Had own workshop Berlin, after war Frankfurt. Since 1959 teaches in Münster. Represented in several exhibitions, gold medal Munich 1958. State prize of North Rhine – Westphalia 1967. With wife has staged fashion shows with jewelry as main feature in Deutscher Werkbund Berlin, Werkkunstschule Berlin, Jahrestagung der Kunsthandwerker Cologne, Handwerksforum Hanover, bei der Gedok Bonn, and Düsseldorf, Aalborg, Copenhagen, Stavanger, Bergen, Trondheim, Oslo. Plate 346

Gilbert, Sir Alfred 1854–1934
Britain

Sculptor and metalworker. Studied Heatherley's and Royal Academy Schools, London, Ecole des Beaux Arts, Paris 1876–8, Rome 1878–84, ARA 1887, RA 1902. Knighted 1932. Most famous sculptural commission was Shaftesbury Memorial Fountain (Eros) in Piccadilly. Apart from the great ceremonial badges, particularly Preston and the Royal Drawing Society, he modelled some notable ceremonial silver tableware, including a table centrepiece made for Queen Victoria's Golden Jubilee 1887, now in the Royal Collection, and a few domestic jewels in iron. A superb gilt ring by him belongs to the Royal Academy. Plate 238

Gill, Eric 1882–1940
Britain

Artist, letterer, engraver, writer. Studied Chichester School of Art, lettered new Medical School at Cambridge, established workshop at Ditchling, Sussex; carved Stations of Cross in Westminster Cathedral and has produced other important sculpture. 1925–39 designed type for Stanley Morison at Monotype Corporation, the basis of much modern typography on silver. Many wood-cut book illustrations and publications on his conversion to Roman Catholicism and on his craft theories. Dunstan Pruden, silversmith of Ditchling, and David Kindersley, letterer of Cambridge, are two of his notable students. Much work for the Worshipful Company of Goldsmiths. Plate 320

Gillgren, Sven Arne 1913–
See GAB pages 30–1 140, plates 53–58 60 62 111 139

Gleadowe, R. M. Y. CVO 1888–1944
Britain

Designer. Worked with silversmiths from about 1929 until death during war, after saga of Stalingrad Sword for which he was awarded a CVO. Worked mainly with H. G. Murphy, Wakely & Wheeler and Barnards. He was art master at Winchester, and Slade Professor at Oxford. Most notable masterpieces, apart from those at Goldsmiths' Hall, were Birmingham Jewelers' Association rosewater dish, given by the Goldsmiths' Company to Birmingham Corporation, Guildhall vases, silver and other works of art for Winchester College and Cathedral. His work was marked by delicate pictorial and in later pieces, when he and Murphy were co-operating, by plainer surfaces with fluting and curved outlines. See page 137, plates 321–3

Goldsmiths and Silversmiths Company Limited
Britain

Retail and manufacturing jewelers. Established 1890 by William Gibson at 112 Regent Street, after failure of the Goldsmiths Alliance in the City. 1898 public company. 1899 registered their mark at Goldsmiths' Hall. Father of Geoffrey Bird of this firm started Searles in the City c. 1904 and two other assistants started Wilson and Gill, incidentally taking with them supply rights of Charles and Richard Comyns whose 'Cherub' presentation set was particularly in demand. 1926 new and larger premises built on same site. Under Eric Hodges, leading patrons of new silver between the wars, often working with Worshipful Company of Goldsmiths, using Harold Stabler to design a notable series of blue enamel boxes. Since 1947 Alex G. Styles has been permanent staff designer. 1952 acquired Garrards, the latter being adopted as the name for both firms. Plate 324

Goodden, RY RDI CBE Professor 1909–
Britain

A nephew of R. M. Y. Gleadowe and an architect who early began to design in silver and other materials and won a sports trophy competition while a student at the Architectural Association. Subsequently chosen by Royal Society of Arts as Royal Designer for Industry and is Vice Principal and Professor of Silver and Glass Design at Royal College of Art. Took a prominent part in South Bank designs during 1951 Festival of Britain, where he was responsible for the Lion and Unicorn building; he was one of the architects of the new Royal College of Art building in Kensington Gore. Designed the Queen's Coronation cup in the collection at Goldsmiths' Hall, London. Member of Court of Worshipful Company of Goldsmiths. Plates 323 369 371

Gorham
USA

Founded by Jabez Gorham in Providence, Rhode Island, 1831. 1841 was joined in firm by son John who, on his father's retirement in 1847, installed first steam engine in factory and broadened scope of business. 1861 New York sales office was opened to cope with sales distribution throughout North America. During American Civil War manufacture of electroplated nickel silver was introduced. 1868 company adopted Sterling Standard and a trade mark of a lion (English hall-mark); an anchor (for Rhode Island); and a 'G' (Gorham). 1865 organized as a corporation. 1873 New York retail sales store opened. 1878 San Francisco and Chicago branch sales offices opened. 1887 Edward Holbrook became fourth president of company and brought over from England William J. Codman as a designer. 1876 to 1915 earned many awards for design and craftsmanship in silver and gold at various world Expositions held during that period. World War I and II made war materials for the US and Allies. Won three times 'Army-Navy Award for High Achievement in the Production of War Materials'. Currently has large research and engineering facilities whose most notable recent development was that of the one-piece sterling knife handle. See Black Starr and Frost.

Grenville, John 1918–
Britain

Artist-craftsman. Studied painting Farnham School of Art 1938–40. Silversmithing Central School of Arts and Crafts, London, 1945–47. Designed jewelry from 1957. Workshop first in London, then with shop in Clare, Suffolk; now in Stowmarket.

Grima, Andrew 1921–
Britain

Jeweler, founded HJ Company 1946, opened own retail shop Jermyn Street London 1966 where he sells Gerald Benney silver. 1966 won Duke of Edinburgh's award for elegant design, the Design Centre's top award, and the Queen's Award for Industry as a pioneering world-wide exporter. Nine sales tours throughout North America in 1966. See *Modern Jewelry* by Graham Hughes. Plates 131–3

Gübelin
Switzerland

Schweizerhofquai 1, Lucerne. Clock and watchmakers, jewelers. Established 1854 by Mauritz Breitschmid (b 1832) in Lucerne; later joined as partner by Jakob Eduard Gübelin, his son-in-law. Moved to present address on lake shore 1903. J. E. Gübelin succeeded in 1919 by son Eduard, who introduced jewelry manufacture to the firm in 1925. New York branch established 1922, followed by others in St Moritz, Geneva, Zürich, Bürgenstock and Paris. Five of the sons of Eduard Gübelin now direct firm, which has exhibited e.g. New York World Fair 1939 and Diamonds-International Award, New York 1958–60. See Widmer

Hamilton & Inches Limited
Britain

87 George Street, Edinburgh 2. Manufacturers and retailers. Established in 1866 by Robert Kirk Inches and his uncle, James Hamilton, at 90 Princes Street, mainly as jewelry retailers; expanded by taking over next door silversmiths. 1887 moved to 87–8 Princes Street incorporating the firm of Robert Bryson & Sons. Royal Warrant given to Robert and still held by the firm. He was appointed 'His Majesty's Clockmaker and Keeper and Dresser of His Majesty's Clocks, Watches and Pendulums in Palaces and Houses in His Ancient Kingdom of Scotland' – a post dating back nearly 400 years, now discontinued. He became Lord Provost of the City of Edinburgh and was knighted in 1915. He was the last Lord Provost to drive daily to the City Chambers in his own carriage and pair. 1902 the firm made and installed the huge clock on the Edinburgh North British station hotel. Edward James Inches, trained in watchmaking in Switzerland and apprenticed in London, joined his father; Campbell, his son, became sole proprietor at the age of 22 and steered the business through the 1939–45 war. His younger brother, Ian Hamilton Inches, acquired the business in 1950, moving to present elegant premises in George Street. Became a limited company in 1965; sells wide range of products from traditional Scottish pieces to modern jewelry from all over the world; modernized plant, with silversmiths, jewelers and watch and clock repairers; special commissions made for almost every country in the world, for instance a monstrance for All Souls Church, Los Angeles, or a reliquary for Dunedin, New Zealand; often until recently designed by Alan Place. Have also made up designs by the sculptor C. d'O. Pilkington Jackson, notably the maces for St Andrew's University Medical School, and in 1966 for the Heriot-Watt Uni-

versity. At present employs 50 people, much the biggest skilled workshop in Scotland. 1966 centenary exhibition.

Handy and Harman
USA

Refiners, bullion dealers. Founded 1867. Held 5 conferences for craftsmen, 1947–9 at Rhode Island School for Design, 1950–1 at Rochester Institute of Technology, School for American Craftsmen. In 1947 their guest teacher was William Bennett of Sheffield, 1948–9 and 1951 Erik Fleming of Stockholm, 1950 Reginald H. Hill of London. Each course attracted some 12 US students from some 45 applicants. The company's Craft Service was discontinued 1953 after giving notable stimulus and encouragement to American craftsmen. See Craver

Hansen, Hans
Denmark

Retail and manufacturing firm established 1906 in Kolding by Hans Hansen. Flatware, holloware and jewelry widely exhibited: represented in museums in Stockholm, Zürich, New York and Copenhagen. 1953 engaged Bent Gabrielsen Pedersen (b 1928 Copenhagen) as chief designer, particularly for jewelry, who had previously worked briefly for Georg Jensen. Gold Medal Milan Triennale 1960, and many distinctions, including Lunning Prize 1964. Karl Gustav Hansen (b 1914 Kolding) is the art director. He trained as silversmith and sculptor, concentrates on flatware and holloware. This firm with its distinguished showroom on Stroget in Copenhagen is one of the half-dozen leading influences in Danish silver design. See pages 11 142

Harding, Neil 1937–
Britain

Artist-craftsman. Studied Birmingham, Royal College of Art. Now lecturer Leicester College of Art, freelance designer for Arthur Price cutlers and others. Plate 103

Harman and Lambert
Britain

177 New Bond Street, London. Retail jewelers. Established 1760 under the title of Lambert of Coventry Street, to trade as dealers in fine old English and foreign silver and especially ecclesiastical plate. Over a period of some 150 years were holders of Royal Warrant. 1897 amalgamated with Harman and Co at present address. Adopted title of Harman and Lambert, held Royal Warrant for the late Queen Mary, who frequently visited these premises. Shop was redesigned in 1960. Now associated with Messrs Packard of Regent Street and with Northern Goldsmiths' Association group.

Hart, George 1882–
Britain

One of the original Guild of Handicrafts formed by

Charles Ashbee, architect and writer, in the East End of London at Essex House in 1888, a leading influence in the revitalization of European handwork. 1902 the Guild moved with all equipment, stock and impedimenta to Campden in Gloucestershire and became the Campden Guild, the nucleus of Cotswold craftsmen who worked in and restored many local old houses, and owned and cultivated local land. It included many crafts, slowly dwindled, wound up in 1907. But George Hart continued to flourish and worked with Huish and Warmington and his son; a universally popular figure, he was responsible for much church plate and other silver. Robert Welch's workshop is now above George Hart's in the same building, the original home of the Guild.

Hartkopf, Paul Günther 1925–
Germany

Manufacturing and retail goldsmith and jeweler. Very smart shop at Berlinerallee 36; half a dozen craftsmen in next room. Head of Düsseldorf Innung (guild), formed to discuss common projects and problems, for which he travels throughout Germany. 1965 Associate of Worshipful Company of Goldsmiths. Plates 127–30 351–53, page 142

Haseler, W. H. & Company Limited
Britain, Birmingham

Manufacturers 1870–1927. Founded 1870 by William Hair Haseler and from 1896 directed by William Rabone Haseler and Frank Haseler. Produced silver and jewels for Liberty's from 1901 with the mark Liberty & Co (Cymric) Ltd, and from 1903 Tudric pewter; Haseler was still occasionally employed by them even after the joint company was dissolved in 1927. Plates 250 252, page 133

Hennell, R. G. & Sons Limited
Britain

4 Southampton Place, London. In 1736 David Hennell (b 1712) set up in business as a goldsmith under the sign of 'The Flower de Lis & Star' in 'Gutter Lane, ye corner of Carey Lane', beside Goldsmiths' Hall. 1756 joined by his son Robert. The business prospered and in 1769 moved to Foster Lane then the centre of the goldsmiths' trade and where the Hall of the Worshipful Company of Goldsmiths stands today. There the Hennells carried on their business, mostly silversmithing, until 1839 when Robert Hennell broke away and set up as a jeweler in 4 Southampton Street (now Southampton Place), Bloomsbury. His sons succeeded him, but in 1906 sold out to a partnership, one partner being A. W. Hardy, hitherto their manager, one of London's leading pearl experts of the time. During the 19th century, pearls, fine gems, and jewelry became their chief interest rather than silver. Charles Bruno joined as designer in 1928; he was succeeded in 1955 by Edward Tuson who had already studied under Bruno, Tuson

leaving 1967 to become manager of Andrew Grima's new shop in Jermyn Street.

Hill, Atholl 1935–
Britain

Studied Glasgow. Design consultants Tandy Halford and Mills; Ministry of Works; now designer for British Rail. Plates 396–7

Hill, Reginald H. 1914–
Britain

After apprenticeship studied at the Central School of Arts and Crafts and held Worshipful Company of Goldsmiths' Scholarships, later becoming instructor in silver and jewelry design at the School. Carried out many important commissions, many of them in recent years made by C. J. Vander Ltd. See Handy and Harman

Hingelberg, Frantz
Denmark, Aarhus

Royal Court jewelers, retail and manufacturing silversmiths. Founded 1897 by Frantz Hingelberg as retailers with separate workshop for gold and silver. Vilhelm, son of Frantz, succeeded as manager. Expanded silversmithing, co-operating throughout his life with distinguished designer Svend Weihrauch. 1943–4 new workshop built, Harald Jensen following Weihrauch as head of workshop and chief designer. Other designers are: Jacob E. Bang, Vagn Åge Hemmingsen, Enling Borup Kristensen, Knud Holst Andersen. 1960 new shop built at Store Torv 3, one of most modern in Denmark. 1963 limited company. 1966 Frantz Jørn Hingelberg, on father's death, became head of firm. Trained as gemmologist, silversmith and businessman (b 1925). Firm's modern sterling silver outstanding; first overseas exhibition Brussels World Fair 1935, followed by many other shows and prizes. 1948 purveyors to Royal Danish Court. See pages 11 142

Hoffmann, Josef 1870–1955
Austria

Architect, graphic artist and designer. Studied painting at Munich Academy and architecture with Otto Wagner in Vienna. Took prominent part in founding of the Vienna Sezession in 1897, and became leader of the new art movement. Work showed ornamentation stressing geometric forms similar to that of Mackintosh in Glasgow, but with his own Viennese style. 1899 appointed Professor of Architecture of the Wiener Kunstgewerbeschule; a brilliant teacher of great influence throughout Central Europe. With Olbrich he designed the Austrian pavilion at the 1900 Paris World Fair. 1903 with Moser and Wärndorfer founded the Wiener Werkstätte. Two important early works in architecture are Sanatorium at Purkersdorf, 1903–14 and the Palais Stoclet, Brussels, 1904–11, for which he also did the interior decoration, furnishings, and gardens. 1912 he founded the Österreichischer Werkbund which he left in 1920 to take the lead of the 'Gruppe Wien' of Deutscher Werkbund. Continued to practise and teach until 1950 and died at the age of 85 in 1955. His last cutlery design for the Solingen cutlers Pott won a gold medal at the Brussels World Fair 1958 and at the Milan Triennale 1960. Austrian commemorative portrait postage stamp 1966. Plates 63 73 143 265 268 269, pages 38 134

Holden, Geoffrey 1916–
Britain

Trained at Birmingham and the Royal College of Art, becoming silver instructor at Tunbridge Wells and later at Brighton College of Art. Author of *The Silversmiths' Craft*.

Hopea, Saara 1925–
Finland

Designer: jewelry, silver, glass. Studied school of industrial art, Helsinki 1943–6. Travelled and exhibited widely, including Milan Triennale 1954, 1957, 1960. Design in Scandinavia, USA, 1952–6; Brussels 1954; Brussels World Fair 1958. Silver Medal, Milan Triennale 1957. Furniture designer for Majander Oy 1946–8. Glassware for Wärtsila-koncernen/Nötsjöe Glassworks 1952–9. Married to Oppi Untracht, American author on metalwork and craftsman.

Howes, B. D. & Son
USA

Retail silversmiths and jewelers. Los Angeles, Pasadena, Newport Beach, Santa Barbara, Goldwater's, Phoenix and Scottsdale, Huntington-Sheraton Hotel, Wilshire Boulevard. Bought Santa Barbara store in 1948 from L. Eaves & Co. by whom it was founded in 1883. First Howes store opened in 1870 in Clinton Iowa, by two brothers; B. D. Howes, their younger brother always fascinated by pearls, later joined them. B. D. and his son Durward moved West and opened the first Howes store in Los Angeles in 1919, since when they have always been reputed for pearls. They maintain a fine collection of coloured pearls there, many of them from the Mississippi River.

Hügler, Jul.
Austria

Freisingerstr., Vienna. Retail and manufacturing jewelers. Bad Gastein, Rio de Janeiro. Founded 1872 by Mr Julius Hügler. During the monarchy furnisher of the Imperial Court. Branches in Cairo, Karlsbad, lost in the Second World War. Harta Jalkotzy-Duc head of workshop, translated Graham Hughes' *Modern Jewelry* for German edition.

Hutton, William & Sons
Britain

Silversmiths, manufacturers of silver, EPNS, and Britannia metal holloware, flatware and cutlery, at West Street, Sheffield and Hanley Street, Birmingham, with an office in Farringdon Road, London. (There was a small factory attached to this office, for the manufacture of *art nouveau* silver.) Founded 1800, by William Hutton (1774–1842) whose main business was a plater in Birmingham. The Ryland family were his partners for part of the time; William Ryland was later an important member of the staff of Elkingtons. He was probably related to Richard Hutton sickle smith of Mosboro and Ridgeway near Sheffield whose family business stayed there 1678–1949. 1807 William registered a mark at Sheffield Assay Office. By the end of the century Sheffield factory was one of biggest in the trade, and Birmingham a small branch works.

William pioneered production of spoons and forks in new white alloy called nickel silver, German silver, 'Argentine', 'British plate', which had been known for long in China under the name Packfong. The composition of this alloy was first published by Fyffe in the Edinburgh *Philosophical Journal* 1822. 1823 the Prussian Society for the Encouragement of Industry offered a gold medal and prize for erection of a factory for Fyffe's alloy; 1824 Dr Gertner opened factory at Schullberg; 1825 another followed in Vienna; by 1829 Percival Norton Johnson, founder of Johnson Matthey, was producing it in Hatton Garden London, and selling to William Hutton.

William Carr Hutton helped in preparation of the 1844 hall-mark statute. William Carr had moved to Sheffield in 1832, and by 1845 employed 11 people there. His brother James acted as agent for the firm in Canada. 1843 William Carr obtained from Elkingtons, a licence for electroplating only one day after the first Sheffield licence was given to John Harrison: the new process resulted from Faraday's propounding his laws of electrolysis in 1833, was patented by Elkingtons in 1840, and quickly displaced the old 'close plating'. William Carr died in 1865, leaving about £16,000 and five sons, all of whom were educated at Vevey in Switzerland. They all joined the firm – James Edward who managed the London business side of the firm died in 1891 and his brother Herbert in 1904.

Huttons became leading *art nouveau* producers, supplying Liberty's of London, Citroën of Amsterdam etc. The Hutton trade mark for electroplate, a group of crossed arrows, was familiar throughout the world. Herbert's nephew, also called Herbert, was joined in 1908 by his brother R. S. Hutton, who brought to the firm a scientific outlook almost unique in these industries. Great-grandson of the founder, he was a director 1908–17: the firm then employed over 800 people, later increased to nearly 1000. Just before 1914 a new process for the manufacture of flatware, invented by Wilzin in Paris, was purchased jointly by Huttons, Walker & Hall, Dixons and Barker Bros. Sheffield Flatware Company was founded and installed 500-ton presses, designed to make 2 tons of spoons and forks weekly. For most of the war the new plant was engaged in making munitions with a maximum output of over 200 tons per week.

R. S. Hutton who was managing director of this new factory from the start, resigned from William Hutton and Sons in 1917 and in 1920–1 organized remarkable time and motion studies on polishing in the highly mechanized Sheffield Flatware Company. From 1921–32 he was first Director of British Non-Ferrous Metals Research Association; 1932 Goldsmiths' Professor of Metallurgy at Cambridge; 1942 Prime Warden of the Worshipful Company of Goldsmiths. He is an active and distinguished campaigner for better technology in industry, for instance he helped to start the Design and Research Centre for the Gold, Silver and Jewelry Industries founded at Goldsmiths' Hall in 1946, and was Council Member City and Guilds of London Institute since 1932 (Chairman 1939–45).

Herbert Hutton too, Managing Director of William Hutton & Sons Ltd from 1907, resigned in 1923, owing to family disagreements, ultimately causing the firm's demise, its goodwill being transferred in 1930 to James Dixon. From 1925 he was manager of the Bristol silver factory Josiah Williams till they were destroyed by German bombers. The Sheffield Flatware Company, having tooled up for large scale production particularly of the 'baguette' pattern for export to Europe, closed down in 1930 when demand failed in the slump. Modern technique pays only if very carefully organized. See Swaffield-Brown, Elkington. Plates 246 249, page 75

International Silver Company
USA

Emerged 1898 from gradual combining of interest between the Meridean Britannia Company (pewter and Britannia ware), William Rogers (who perfected the silverplating process in the USA) and group of independent silverplate makers. Later in 1898 Holmes and Edwards joined the company. In addition to domestic sterling and silverplated cutlery and holloware, the company is large supplier to America's hotels, restaurants, airlines, etc. Sponsored world flatware competition organized by Museum of Modern Crafts, New York, to celebrate 60th anniversary 1960, won by Sven Arne Gillgren of Stockholm. Twenty-two award winning designs selected from total entry of 206 from 17 countries. Designers thus chosen included Ainar Axelsson of Stockholm, Max Bill of Zürich, Leslie Durbin of London, David Gumbel of Israel, Arthur Pulos of Syracuse New York, Olaf Skoogfors of Philadelphia, Robert von Neumann of Champaign Illinois, and Tapio Wirkkala of Helsinki. Final prize jury included Edward D. Stone architect and Walter Dorwin Teague industrial designer. Exhibition circulated by Smithsonian Institution.

Jaccard Jewelry Company
USA

Founded 1829 in St Louis, Missouri by Louis Jaccard, a French-Swiss immigrant. Business increased rapidly, making jewelry for the increasingly prosperous citizens around St Louis. 1837 Louis was joined by Eugène Jaccard, from Switzerland, and firm became known as E. Jaccard. 1847 they, in turn, were joined by D. C. Jaccard and A. S. Mermod but latter two split away from E. Jaccard in 1864 to establish own firm in St Louis, which was incorporated as Mermod & Jaccard Jewelry Company in 1883 and which, a few years later, bought out the firm of E. Jaccard. 1888, Walter M. and Eugene G. E. Jaccard, sons of D. C. Jaccard went to Kansas City to establish a jewelry store. 1902 Ernest A. Jaccard left St Louis business to join Walter M. in Kansas City. He was last of Jaccards to be actively involved with St Louis business and in 1917 the family sold their interest in that firm. Jaccard store has continued to flourish in Kansas City. Around the turn of the century a lot of gold and silver commemorative pieces and new jewelry was made in the workshop. Now the firm is mainly a retailer of jewelry and silverware, with a workshop for repair work and a small amount of creative work. Jaccard family is still active in the business.

Jensen, Georg
Denmark

Leading firm silversmiths and stainless steel manufacturers with continuously progressive artistic policy and world-wide exports. Established Copenhagen by Georg Jensen (1866–1935), now has branches in London, Liverpool, Edinburgh, Glasgow, New York, Toronto, Detroit, Paris, Brussels, The Hague, Amsterdam, Stockholm, Geneva, Zürich, Düsseldorf, Sydney, Melbourne, Adelaide, Hong Kong, Barcelona. Plates 141–2 150–6, pages 9–12 53 74 93–5 132 134 137 140

Jenson, Søren Georg 1917–

Son of Georg Jensen. 1946 Gold Medal, Royal Academy, 1960 Gold Medal, Milan Triennale. See pages 9–12, plate 19

Johnson, Matthey & Company Limited
Britain

Refiners, smelters, bullion dealers. Founded 1817 by Percival Norton Johnson, pioneered platinum research after 1851. When Johnson took his young apprentice, George Matthey, into partnership in 1860 he retired and firm came under control of George Matthey, John Scudamore Sellon, Johnson's nephew, and Edward Matthey, George's younger brother. All three had scientific qualifications and firm's main business was gold refining, melting, smelting and research. Now the world's largest bullion firm, has branches in many countries, and in addition to its original products, makes component parts for silver and jewelry trades, such as pencil-cases and earring clips. But bulk of business since 1945 has been supplying pure metals for new electrical and electronic industries and for banks. The Chairman 1957–60, George Matthey, is descended from founder's apprentice, and was Prime Warden of the Worshipful Company of Goldsmiths. The Sheffield Smelting Co, the firm's chief competitors, had a Methodist management and opposed the Boer War so hotly as to become unpopular, thus making room for Johnson, Matthey's subsidiary, Oakes Turner. Factories now in Sheffield, Birmingham, Shepherds Bush and Hatton Garden, London, Harlow.

Jones, A. Edward, Ltd
Britain, Birmingham

Founded 1902 by A. Edward Jones (1879–1954). Since 1780, the Jones family were blacksmiths at the hamlet of Elmley Lovett nr. Droitwich. In the 1860s William Jones moved to Birmingham and started a blacksmith's and coach building business in Hurst Street. His son, A. Edward Jones studied the Central School of Art, Birmingham under Edward Taylor and was apprenticed at Woodwards of Paradise Street, Birmingham; specialized in art nouveau; often illustrated in the magazine 'Studio' c. 1906. Emphasis on church work. In 1905 he bought the good will and trade mark, St Dunstan beating a copper bowl, of Llewellyn Rathbone, who became the head of Sir John Cass College in London. Jones' son Major A. K. Crisp Jones now head of firm.

Jünger, Hermann 1928–
Germany

Goldsmith silversmith and jeweler. Studied Munich, workshop Taufkirchen nearby 1958. Bavarian state prize 1963 at Munich International Handicrafts Fair, commendation 1963 and 1966 at the Stuttgart LGA Museum international handcrafts exhibition. Employs three craftsmen. Many commissions for church work including cross for exhibition by German government at Montreal World Fair 1967. Also makes jewels as free artistic expressions. Has highly personal style of rich and random ornament reminiscent of drawings by Paul Klee. One-man shows Munich 1965–6, Bielefeld, Kausbeuren 1967. Plates 423 431–2, page 142

Kirk, Samuel & Son Inc.
USA

Manufacturers and retailers, silversmiths and jewelers. Founded by Samuel Kirk who opened his shop in Baltimore, Maryland in 1815; till 1820 traded as Kirk & Smith, thereafter as Samuel Kirk, descendant of the English 17th century silversmith Jonah Kirk registered 1696, and Sir Francis Child, Lord Mayor of London 1669. Patrons included General Lafayette, Robert E. Lee, Jeferson Davis, the Lowells, the Roosevelts and the Astors. In 1828 Samuel introduced Kirk Repoussé, the flower and foliage design, sometimes still called 'Baltimore Silver'.

Survived five buildings, two fires, seven wars, five panics and eleven major depressions. Kirk agencies in every State of USA as well as Europe, South America and the Far East. Members of the Sterling Silversmiths' Guild with Gorham, Alvin cutlers, International, Lunt in Greenfield Mass., Oneida, Reed & Barton, Tiffany and Towle.

Knudsen, Anni and Bent 1927– 1924–
Denmark

Own workshop 1956. Artist-craftsmen, exhibited extensively in USA, Scandinavia and Germany, both through Den Permanente, Copenhagen, and on their own. Employ a dozen craftsmen; have their showroom in a wartime concrete bunker in their garden at Kolding.

Koppel, Henning 1918–
Denmark

Sculptor, silver designer. Studied Øregaard School, Bizzie Høyer's school of draughtsmanship, and Danish Royal Academy of Art under Professor Utzon Franck, intending to become a sculptor. 1938 went to Paris to Académie Rançon where sculptor Malfrey was his teacher. Married, moved to Stockholm where he worked with a stonemason and also painted watercolours. Began designing jewelry at the suggestion of 'Svenskt Tenn', shop through which he had been selling his paintings. Returned from Sweden in 1945 and began association with Georg Jensen. Won many awards. Works out silver ideas in clay model form first; inspired by nature – the way a branch grows from the tree trunk, for instance. Feels indebted to Harald Nielsen, the silversmith at Jensen's, for his valuable criticism and advice. See pages 12 142, plates 22–31

Korsmo, Grete Prytz 1917–
Norway

Studied State School of Arts and Crafts, Oslo, and Institute of Design Chicago. Apprenticeship as goldsmith. Practised as goldsmith in own workshop and worked with Tostrup, Oslo. Travelled extensively. Exhibited in Norway and abroad, notably at the Triennale, Milan, and Formes Scandinaves, Paris. Personal exhibition at Skansehallen, Stockholm, at Artek, Helsinki, and at the Biennale. Grand Prix and Gold Medal at the Triennale 1954, 1957; Lunning Prize 1953. See Tostrup.

Lacloche
France

8 Place Vendôme, Paris. Retail and manufacturing jewelers and art gallery. Created 1897 by brothers Fernand, Jules, Leopold and Jacques Lacloche. Leaders of fashion during 1920s for enamels and carved stones in the oriental manner, and for tiny geometrical patterns in precious jewels.

Lalique, René Jules 1860–1945
France

Jeweler, silversmith, glass-maker, artist and decorative designer; greatest French jeweler of the *art nouveau* period. Apprenticed to Louis Aucoc, Parisian jeweler and silversmith, 1876, also studied at the École des Arts Décoratifs. 1878–80 studied at an art school in London. By 1885 had established his own jewelry workshop in Paris. Represented anonymously at the Paris World Fair of 1889 and awarded a prize. First known at the Salon des Champs de Mars in 1895; 1897 Chevalier Légion d'Honneur; represented at the opening of S. Bing's shop 'la Maison de l'art nouveau', and his work was acclaimed with enthusiasm at the Paris Universal Exposition 1900. Exhibited at Turin exhibition 1902; one-man shows in London 1903 and 1905. Opened Place Vendôme shop 1895, and in 1903 designed and built his shop at 40 Cours Albert I. In 1909 he leased and in 1910 bought a glass factory at Combes-la-Ville. Pages 9 135–6

Lessons, Kenneth William 1933–
Britain

Artist-craftsman. Studied Sheffield and Royal College of Art. Own workshop 1957. See page 77

Lettré, Emil 1876–1954
Germany

Artist-craftsman. Worked Vienna, Budapest and Paris before moving to Berlin. From 1933 Director of the Staatliche Akademie in Hanau. Grand Prix Paris 1937. Taught H. G. Murphy in Berlin. See page 138

Leysen, Frères
Belgium, Brussels

Rue de Laeken and Rue Royale, 28 rue Marché-aux-Poulets; 53 Bd de Waterloo. Retail jewelers and silversmiths. Founded 1855 by great-grandfather of present managers, the brothers Pierre and Jacques Leysen. 1920 their father Henri built the present typical family jeweler's shop, in Louis XVI style, as the headquarters.

Liberty & Co.
Britain, London

Started 1875 by Arthur Lasenby Liberty when he left firm of Farmer & Rogers, importers of Oriental *objets d'art*, which closed in 1874. He commissioned a series of 'Art fabrics' which made firm famous in England and on Continent, where he gradually built up a chain of branches. The work was rather light and undistinguished but acquired a good reputation. The fabrics gave the local name 'Stile Liberty' to *art nouveau* in Italy. The first silver mark entered at Goldsmiths' Hall 1894. Liberty used many metalwork designers, his chief suppliers probably being William Hutton & Sons of Sheffield and W. H. Haseler & Co of Birmingham; 1899 he launched Cymric range of silver and jewels, distinguished by new hall-mark Ly & Co, and Tudric pewter followed, in 1901. Liberty's did not disclose the names of their designers but amongst others the following worked for them in their early youth: Oliver Baker (1856–1939), A. H. Jones, Bernard Cuzner (1877–1956), Reginald (Rex) Silver (1879–1965), Arthur Gaskin (1862–1928), Archibald (Charles?) Knox and Jessie M. King. Some of the *art nouveau* pieces were still sold as late as the 1920s and even (exceptionally) the 1930s. A successful marriage of art and commerce, the Cymric company wound up in 1927. See Shirley Bury *Architectural Review* February 1963, and in *British Decorative Arts of the late 19th Century in the Nordenfjeldske Kunstindustrimuseum,* Trondheim 1961–2. Plates 248 250–2, page 131–3

Linz
USA

Retail silversmiths and jewelers, manufacturing jewelers, Dallas, Texas. Founded 1877 Denison, Texas, by Joseph Linz, watchmaker's assistant, in same year as last Indian raid in Texas. 1879 moved with brother Simon to Sherman, where building still stands with Linz name carved on it. 1891 moved to Dallas now with brothers Albert and Ben. Simon launched Linz award given annually through *Dallas Times Herald,* to local citizen who best serves city with remuneration. 1899 third move in Dallas to new seven-storey Linz building, then the tallest fire-proof structure in the South, now called the Rio Grande Building. Albert Linz at 84 printed a card with his favourite motto 'Don't wait for the funeral, send me the flowers now'. He used to sell jewelry in the country renting a surrey and horse from Sherman. 1962, on 85th anniversary in the Main Street, Dallas shop to which they had moved, exhibition staged of Patek Philippe watches; and four centuries of English silver from the Worshipful Company of Goldsmiths of the City of London called 'Historic and modern treasure selected from the finest British handmade gold and silver of the past 400 years from Goldsmiths' Hall'. 1964 sold to the second largest jewelry operation in USA: Gordon Jewelry Corporation headquarters in Houston. Joe Linz, the last of the family in the firm, retired.

Littmark, Barbro
See Bolin, page 142, plates 41–2

Macdonald, Frances 1874–1921
Britain

Scottish decorative designer and metalworker. Younger sister of Margaret Macdonald. Educated at Glasgow School of Art. Worked both alone and with her sister. Married the architect J. Herbert McNair in 1899 and afterwards collaborated with him in design of furniture and stained glass. Taught enamelling, gold, silver and metalwork at Glasgow School of Art from 1907. Page 133

Macdonald, Margaret 1865–1933
Britain

Studied Glasgow School of Art, designed and made many works in metal, usually with her sister Frances. Married C. R. Mackintosh 1900, and then collaborated in all his work; considerably influenced development of his style. Her later work includes gesso panels and stained glass. See page 133

Mackintosh, Charles Rennie 1868–1928
Britain

Architect and designer. Most original British artist of his day and leader of 'Glasgow School', pioneering group of architects and designers who exercised great influence in England, on the Continent and in America. Studied Glasgow School of Art 1885, then part-time student while apprenticed to Glasgow architect, John Hutchinson. Joined firm of Honeyman & Keppie 1889. Series of awards including travelling scholarship to France and Italy 1890. Won competition for new Glasgow School of Art building 1897. Designed, decorated and furnished series of tea rooms in Glasgow for Miss Cranston 1897–1910. Designed Scottish Pavilion, Turin Exhibition 1902. Married Margaret Macdonald. Designed tableware prototypes and architectural metalwork. Plates 240–3, pages 9 56 133 135

Magnussen, Eric 1884–1961
Denmark

Artist-craftsman, self taught and at Kunstgewerbeschule, Berlin 1907–8. Worked for International Silver, USA, *c.* 1938–9. Own firm 1901–33 and 1939–60. First exhibition in Copenhagen *c.* 1901, exhibitions *c.* 70 USA cities during the 1930s. One of Danish silversmith pioneers of *art nouveau.*

Mainar, Montserrat 1928–
Spain

Artist-craftsman. Studied Escuela Massana, Barcelona, specializes in enamel with silver, ceramics. Worked for Monastery of Monserrat: made frontal of the main altar, chalices, etc., 'Gran Premio' of enamels in the third Hispano American Biennale, 1955. One-man exhibitions Barcelona 1956 and 1960. Own workshop Barcelona.

Malinowski, Arno 1899–
Denmark

Trained as designer and engraver in Copenhagen. Designed for Jensens. See page 11

Mappin & Webb Limited
Britain

Silversmiths. Joseph Mappin worked as engraver in Fargate Sheffield 1797–1817, and originated firm. 1908 it became a public company, John Newton Mappin having deserted his family firm of cutlers Mappin Bros, which he subsequently bought up to found a new firm with George Webb, architect, as partner. Acquired Heeley Rolling Mills, high quality foundry started in an old aluminium works 1896. 1913 founded the Sheffield Silver Plate and Cutlery Company for cheaper production; 1951 this had some 300 employees and made some 600 gross pieces of cutlery each week. Now part of British Silverware Ltd. Plates 366–7, page 75

Marks, Gilbert 1861–1905
Britain

A leading silversmith in the arts and crafts movement; his heavy hand-made pieces were very popular in the 1890s. Plate 239

Martinazzi, Bruno 1923–
Italy

Sculptor, silversmith, jeweler. Made silver 1957–9. Makes everything himself in workshop, Turin. One-man show 1964 Goldsmiths' Hall London, Geneva, Munich and Turin. See page 142

Masriera, Luis 1872–1958
Spain

Belonged to well-known family of painters and jewelers. His grandfather founded manufacturing firm in 1839, named 'Masriera e Hijos' 1872, 'Masriera Hermanos' 1886. Studied at School of Arts, Geneva, under famous enameller Lossier. His pioneering work revitalized jewelry design, first in Barcelona, then through Spain and South America from about 1895. As jeweler won 1908 Grand Prix at the International Exhibition of Zaragoza and 1913 at Ghent. As theatre designer won 1925 Grand Prix at Paris. Exhibited widely. 1915 associated with Carreras, established 1750. The firm Masriera y Carreras SA now managed by Sr Juan Masriera, son of Luis.

Meister

Switzerland, Zürich

Branches at 28 Bahnhofstrasse and in Paradeplatz. Manufacturing and retail jewelers and silversmiths. The Meister family traces its history to 15th-century Schaffhausen. Johann Jakob Meister (1815–1900) was a doctor and tutor at Zürich University, and four of his sons were chemist, merchant, engineer and dentist. Emil, the fifth (1847–1921), founded the firm. After goldsmith's training in the Horgen workshop of J. J. Hess he worked for five years in Boston and New York, London, Paris, Belgium, Germany and Geneva, always making sketches in his pocket-book for his jewels. 1874 he returned home, married the daughter of Mr Hess, and soon after inherited the firm, sharing it with his brother-in-law Hess. They quarrelled; Emil bravely opened own workshop with shop window on corner of Münsterhof/Storchengasse in 1881. Wife died 1884, and the same day

he had to pay 16,000 francs for a law case in Horgen in which his brother-in-law had involved him. Had to move to smaller shop at 8 Post Street, and here at last he succeeded. 1897 moved to the fashionable Hotel Baur en Ville. 1909 he entrusted whole business to his second son, Eduard Meister (1880–1954), who had studied in Geneva at Giron et Lamunière for four years, in Paris for two years, and in London and Munich. Altogether he was abroad for eight years. 1912 he married Emmy Wydler; 1925 shop was burgled, lost nearly one million francs worth of stock. Thief was caught in Vienna after seven months, but only one third of stock recovered. Eduard designed more, met customers less.

His son, Walter Meister (b 1917), took over firm in 1940, having studied in Switzerland and Germany at Hanau and Pforzheim, and visited North and South America. Co-founder of Swiss Gemmological Society, its president for three years. Supervised candelabrum made in the firm's workshop and given to Worshipful Company of Goldsmiths for their collection, after British Week in Zürich 1963. Plate 358

Mellor, David 1930–
Britain

Studied Sheffield College of Art, Royal College of Art London. Consultant designer to Walker & Hall (silver) and also to engineering firms (street lighting, calculating machines, etc.) and British post office. Important commissions include: Ministry of Works cutlery for the British embassies. Large candelabrum for the City of Sheffield. Ten candelabra for the Shell Group to use in Directors' dining room at new London office. Awarded RDI. Staffed Worshipful Company's exhibitions in Stockholm 1962 and Zürich 1963. Several Design Centre awards. Own workshop Eyre St, Sheffield 1954, 1 Park Lane 1960. Plates 122 384–89 424–5, pages 139 140 142 end paper

Michelsen, Anton
Denmark

Bredgade, 11, Copenhagen. Royal court jeweler. Retail and manufacturing firm. Michelsen family were blacksmiths on island of Funen; firm's founder, Anton Michelsen (b 1809), was apprenticed to a goldsmith in Odense, then pursuing his training in Copenhagen, Berlin under the royal goldsmith Georg Hossauer, and Paris under Mention et Wagner. Set up as a goldsmith in Copenhagen 1841. The royal court soon encouraged him; 1848 made Michelsens their Goldsmiths and Insignia Jewelers and commissioned from him, for instance, the box for the seal for the new 1849 constitution, the gold cup set with diamonds in 1850 for Countess Danner, and various decorations and insignia. 1855 he first exhibited in Paris. 1877 he died and was succeeded by son Carl. 1878 firm made toilet set for the wedding of Princess Thyra to the Duke of Cumberland, a replica of an Augsburg rococo set at Rosenborg Castle, and in 1892 the largest piece of silver ever made in Denmark, three epergnes, one of them 6 ft 3 in. long and two candelabra, given by the country gentlemen of Denmark as a golden wedding present to King Christian IX and Queen Louise. 1888 N. G. Henriksen, sculptor, appointed master of the workshop, designing silver objects in a naturalistic manner; Harald Slott-Møller the painter designed narrative pieces; Th. Bindesbøll introduced modern style, further stimulated by Georg Jensen who worked for Michelsen 1892–4.

Carl died 1921, being succeeded by Poul (1881–1957), for whom the fine architect Kay Fisker designed a few outstanding pieces in 1927, represented in Copenhagen Kunstindustrimuseum. Ib Lunding experimented with decorated surfaces. Firm held a centenary competition 1941, leading to collaboration with young architects Tove and Edv. Kindt Larsen, and with Ole Hagen, Preben Hansen, and Flemming Lassen. Centenary exhibition at Copenhagen Museum 1941. Since then designs commissioned from Nanna Ditzel and Arje Griegst. Most important single artistic event in firm's career was decision to produce Arne Jacobsen's famous stainless steel cutlery, initially at his invitation for his Royal SAS Hotel in Copenhagen in 1962, where it was unsuccessful and withdrawn, though it has subsequently developed lively world-wide sales. Jørgen Michelsen now head of firm. For pictures of 125th anniversary staff celebrations see *Guldsmedebladet* 8/1966. Plates 123 310–13 339–40 437, pages 11 142

Møller, Inger 1886–1966
Denmark

Went to Copenhagen 1909 as apprentice to Georg Jensen (her cousin Thorolf Møller married a sister of Georg Jensen's wife and became manager of the firm Georg Jensen and Wendel), made trips to London and Paris, and stayed with Jensen until 1921, when she went on scholarship to Italy. Since 1922 had her own workshop in Copenhagen with no employees, often having her work made up elsewhere though always finishing it herself; has never designed for other firms and only sells her work through Den Permanente and the Danish Handcraft Guild. Kept individual style 'in the school of Jensen'. Exhibited 1925 in Danish Pavilion in the Paris World Fair. Decorates her work by sawing out and welding on; in a number of pieces ordered by the Tuborg Foundation for the New York World Fair in 1939, she welded various borders on bowls and match-boxes. Greatest efforts devoted to shaping and forming holloware; made silver coffee pot for King and Queen of Denmark, 1935, on the occasion of their wedding. Always remained a craftsman, with work in many museums and collections, but no designs in mass-production. One-man show Copenhagen Kunstindustrimuseum 1966. Plates 150, 341. Pages 72, 94.

Moritz, Andreas 1901–
Germany

Studied London Central School of Arts and Crafts.

Sculptor and artist-craftsman, teaches Nuremberg since 1952, medal Milan Triennale 1954, gold medal Munich 1958, one-man show Munich 1961, distinction Brussels World Fair 1958. Leader of a classical, restrained German style. Plate 350. Page 71.

Morris, May 1862–1938, Britain

Designer, jeweler, embroideress. Daughter of William Morris who trained her. Designed for Morris & Co, and took over embroidery section of the firm 1885. Lectured in England and in USA 1910. One of the founders of the Women's Guild of Arts 1907 and for some time its chairman.

Morris, Talwin 1865–1911
Britain

Designer and member of Glasgow School. Trained as architect. Later worked on staff of the periodical *Black & White*. Became art director for Blackie & Sons 1890. A prolific designer of bookbindings; also furniture, stained glass and metalwork.

Mucha, Alphons Marie 1860–1939
Czechoslovakia

Painter, graphic designer and decorative artist. Studied Munich, Vienna and Paris (principally Académie Julian) 1890–4. Friend of Gauguin, Strindberg and Rodin. Started own design studio with Whistler. Rose to fame thereafter for his posters and decors for Sarah Bernhardt, following six-year contract with her from 1894. Also active as book designer and illustrator. Contributed to *La Plume* and other magazines. Also did wall paintings for the South-Slavic pavilion at Universal Exposition in Paris 1900, and later for theatres and public buildings in Berlin and Prague. 1900 he brought out his *Documents Décoratifs* which, with its designs for furniture, jewelry, lace and household objects, was recognized as the authoritative statement of the aims of *art nouveau*. Designed jewelry for Georges Fouquet 1898–1905 and his rue Royale shop. Exhibitions at Grosvenor and at Arthur Jeffress Galleries, and Victoria and Albert Museum, London 1963. After World War I recognized as leading Czech artist and commissioned to design coat of arms, stamps, and banknotes for the new state. 1920 exhibited his pictures of 'The Slav Epic' in USA. See *Mucha* by Brian Reade. Pages 135–6.

Murphy, H. G. 1884–1939
Britain

Artist-craftsman, teacher. Studied at Central School of Arts and Crafts, with Henry Wilson, and then in Berlin with Emil Lettré. 1913 set up workshop in London. Taught from 1907 at Royal College of Art, London, then at Central School, becoming first silversmith to be appointed Principal. Plates 320–1. Page 137.

Murray, Keith 1892–
New Zealand

Architect and designer. Came to Britain from New Zealand at top of his form as an industrial designer. For pottery he was retained by Wedgwood, for glass by Stevens & Williams and, at the time of Royal Academy exhibition 1935, a number of his silver designs were made up by Mappin & Webb and exhibited. They issued special catalogue of his work. Served on most committees dealing with industrial art in London, and later planned Wedgwoods new factory at Etruria, going back to his architectural career and leaving industrial product design behind.

Nayler Bros
Britain

45 Broadwick Street, Golden Square, London. Founded 1870 by grandfather of wife of present manager C. T. Smith. The designer, H. Mills, was a partner and a regular exhibitor in the Art Council Demonstrations at Goldsmiths' Hall, London. Plates 375–6

Nielsen, Harald
See pages 94–137, plates 6 8 11–13 16 17

Nilsson, Wiwen 1897–
Sweden

Artist-craftsman, silversmith and jeweler. Studied in the Lund workshop of his father Anders Nilsson, then in Hanau 1913–14 and 1920–1 and Copenhagen. Worked in Paris studio of Georg Jensen, while studying in Grande Chaumière and Colarossi Academies. Own workshop in Lund since 1927. 1928 took over his father's business. Many exhibitions, including National Museum Stockholm 1959. Leader of geometrical style of 1920s and 1930s, particularly admired and respected in Sweden. Gold medals Paris 1925, Milan Triennale 1951; the first Gregor Paulsson award of the Swedish design society (Slöjdföreningen) 1955, gold medal Swedish Goldsmiths' Society 1956, Prince Eugen medal 1958. See pages 11 137, plates 337–8

Ohlsson, Olle 1928–
Sweden

Stockholm Old City (Gamla Stan). At 15 apprenticed to Hallberg jewelers, qualified fully 1949. Worked for Erik Fleming at Atelier Borgila. 1954 evening classes at Konstfack. Designer for two years at Gekå jewels. Spectacular one-man exhibition at Nordiska Company, Stockholm 1966, the first occasion on which his work appeared in public. The first Scandinavian designer/craftsman to evolve personal textured surface, an idea first investigated by Gerald Benney in London for a chalice at Goldsmiths' Hall 1957. Since 1965 freelance artist with workshop at Svartensgatan 5 Stockholm. Plates 99 100

Olbrich, Josef 1867–1908
Austria

Architect, graphic designer. 1890–3 studied Vienna, 1893 visited Italy Tunis France, 1894–9 worked with Otto Wagner. 1897 co-founder of Sezession, designer of the Sezession building in Vienna 1898. Invited to Darmstadt 1899 by the Grand Duke of Hesse where he designed main buildings in the duke's artists' colony there Theresienhöhe. 1903 co-founder of the Bund Deutscher Architekten (BDA). Plates 262–3

Oneida Ltd
USA, Oneida, New York

Wallingford (1877–80) and Sherrill USA; Niagara Falls, Canada (established 1916); Bangor, Northern Ireland; Toluca, Mexico. Cutlers and silversmiths. Perhaps the most extraordinary history in the industry. The Oneida Community founded 1848 by John Humphrey Noyes on Oneida Creek near Oneida Lake in New York State, USA, in Jonathan Burt's sawmill. Noyes' basic principle was the abolition of selfishness and private property, including wives. His wish was for spiritual equality and concentration on breeding the best type of child which would be brought up not by its own parents but by the whole community. As George Bernard Shaw said, this was the only attempt by mankind to breed the 'Superman', and, 'The question as to what sort of man they should strive to breed being settled by the obvious desirability of breeding another Noyes'. Noyes believed that the second coming of Christ had already occurred during the first century AD, and that the millennium had already arrived.
In 1849 membership of the community was 87, in 1850 it was 172, in 1851, 205. 1849–60 the 'Old Mansion House' where some descendants of the community still live, was built. Slowly agricultural business yielded to more profitable metalwork, particularly steel animal traps invented by Sewell Newhouse, a trapper living in the nearby village Oneida Castle. By 1860 these traps were standard throughout USA and Canada; for 70 years the Hudson's Bay Company used them exclusively. In 1865 the community undertook silk, and in 1877 the new Wallingford branch started its own industry, knives, spoons and forks. In 1880 family ties became too strong within the community; families were recognized, and the spirit of self-abnegation weakened. In 1881 the Oneida Community became a joint stock company and they still speak of their history as 'before joint stock' and 'after joint stock'. In 1886 the founder died and his son Pierrepont Noyes (b 1870) soon became prominent, first running his own firm in New York, then joining the community in 1895, leaving it for other public activities in 1917, returning to active participation in it in 1930.
In 1913 silver manufacture was moved from Niagara Falls to Sherrill; by 1950 Oneida Community sales were three times what they had been in 1927, giving a surplus of $11 million in addition to wages bonuses of $9 million. In 1950, at the age of 80, Noyes retired. The Sherrill

plant is now probably the biggest in USA, employing as many as 100 travelling salesmen, 2,933 people.

Community Plate, Oneida's first product, was introduced to Britain in 1919 as an import, then manufactured in 1926 by the Oneida subsidiary Kenwood Silver Company in Sheffield. In 1960 the Sheffield factory closed, a large new building at Bangor, County Down, Northern Ireland, taking its place. There are now 350 employees there, making three grades of silver plate flatware, using the 'Balanced Plating Process' to put most of the silver deposit in places where there is most wear – on the back of each piece, sometimes with an overlay of solid silver; and four different qualities of stainless steel. The factory is one of the biggest and most highly automated in the country.

The Canastota, New York, plant made bus bodies. During World War II it made mobile photo laboratories. It was doing poorly, and Oneida took it on at the US Government's request. After the war, it was reconverted to the building of school bus bodies. Sold 1952. Oneida now have over 40 trademarks, e.g. Heirloom Sterling, Community Stainless, Oneidacraft. Perhaps the world's biggest maker of good stainless flatware.

Osman, Louis 1914–
Britain

Architect, artist, designer. Studied Bartlett School of Architecture and Slade School of Art. Established architectural practice 1936. Since the war designed furniture and silver, including an altar set commissioned by Worshipful Company of Goldsmiths, and projects at Lincoln, Exeter and Ely cathedrals. Occasional jewels. Won Topham Trophy competition 1965 and two prizes De Beers jewelry competitions at Goldsmiths' Hall 1961. Plates 411–17 422, page 142

Oved, Sah 1900–
Britain

1923 worked under J. Paul Cooper. 1924 own workshop. 1927 on, worked 'Cameo Corner', Museum Street, London, specializing in Jewish plate and antique jewelry with her late husband Mosheh Oved. 1934–47 shop in Jerusalem. 1953 wrote *Book of Necklaces*. 1965 married architect Hugh Hughes.

Padgett & Braham Limited
Britain

Founded 1870 as C. and W. Padgett by the father and uncle of C. S. Padgett, the present principal of the firm. Well known London box-makers before the war when they absorbed the business of Dumenil. Subsequently they expanded into all forms of enamel work, making many of the official badges of office, decorations and insignia, and specializing in fine commissioned work by young designers. Took over Wakely & Wheeler on Arthur Wakely's death; and Stanley Padgett Prime Warden of the Worshipful Company of Goldsmiths 1957, did much to keep the standard of design and crafts-

manship high. July 1966 the firm employed 90 people with five firms in the group. W. H. Manton smallworkers of Birmingham, Pairpoint gilders, Flutemakers Guild, makers of silver flutes, Wakely and Wheeler and Monarch Shield for protective lacquering, together perhaps constituting the biggest range of skilled craftsmanship in the world.

Persson, Sigurd 1914–
Sweden

See pages 93 97–8 141–2, plates 137 167–87

Pinches, John Limited
Britain

Medallists. Founded 1840 by four brothers who came to London from Shropshire to study die sinking and medal design; they were descended from William, son of Thomas Pinches, Parish Clerk of Acton Burnell in Shropshire. He had gone to Birmingham to start plating business, at the end of 18th century, 'banking' his customers' plate as well as repairing it. Present management includes J. R. Pinches, Senior Director, trained at the Slade school, John Harvey Pinches his son, Leslie Ernest Pinches, Leonard Richard Pinches. 1958 moved to present factory in Clapham from factory at 21 Albert Embankment (opposite Tate Gallery) which had been their home for 60 years. Notable medals include coronation of Queen Victoria and subsequent monarchs, Royal Geographical Society's celebration of Shackleton's Antarctic voyage 1907–9, Distinguished Flying Cross 1918 and Olympic Games 1948. Medals made in 1851 Great Exhibition, Crystal Palace were struck in a hammer press, three men being needed to turn the hammer bar. Similar presses are used today but with power drive. In 1945 manpower had sunk to six men, but by 1965 had increased to 120.

Peter George Wyon (b 1710) an engraver, founded another famous medal firm patronized by George II. Two of his grandchildren and a great-grandson were chief engravers at the Royal Mint, the best perhaps being William Wyon, designer of George IV's last coinage and Victoria's bun penny, and the first postage stamp, the embossed penny black. Pinches made the Wyon medals from 1905, finally absorbed their business in 1933.

Pinton, Mario 1919–
Italy

Artist-craftsman. Studied Venice, Monza, Milan (under Marino Marini). Teaches goldsmithing and sculpture, Padua. Own workshop 1954 Padua. Seven one-man shows Italy and Germany, in Munich 1964 for the Bavarian government. Bronze Medal Milan Triennale 1954; Gold Medal 1957.

Place, Alan 1927–
Scotland

Silversmith and jeweler, studied at Leeds and Edinburgh, taught at Edinburgh. Designed for Hamilton and Inches; produced notable set of silver for London borough of Twickenham and plate for Sidney Sussex and Caius College Cambridge. 1966 opened new workshop at Kinross, forming retail and manufacturing partnership Place and Reid of St Andrews. Made plate for Stirling University and for Leeds University given by the Worshipful Company of Goldsmiths.

Plata Lappas
Buenos Aires

Founded 1887 by Alcibiades Lappas, a Greek immigrant. Awarded high diploma of honour, and gold medal at the exhibition, celebrating the 100th anniversary of independence in Argentina, in 1910. Also awarded the Grand Prix of the Town Exhibition of Industrial Arts and the Grand Prix and Gold Medal of the Rosario Exhibition. 1925 Grand Prix and Gold Medal, Bolivar Exhibition; 1926 Grand Prix and Gold Medal, Philadelphia Exhibition; 1930 Grand Prix d'Honneur and Gold Medal, Seville Exhibition. In 1938 the founder died but the firm has passed on to members of his family.

Plata Meneses
Spain, Madrid

Founded 1840 by M. Leoncio Meneses and continued as a family firm until 1943 when it became Meneses Goldsmiths SA. Still controlled by the family, who own almost all the shares. Mr Fernando Meneses, Marquis of San Julian de Buenavista, is Chairman and Managing Director. The factory produces wide range of electroplated nickel silver flatware and holloware and has a special department manufacturing church silverware, such as chalices reliquaries, thrones, etc. The sales organization currently covers Spain, Mexico and Latin America.

Popham, Philip 1919–
Britain

Student and subsequently an instructor at Royal College of Art, for whose collection he has designed and made table plate.

Pott, Carl 1906–
Germany

Cutlers. Founded 1904. See pages 9 38 73, plates 63–76

Prado, Nilda Nunez del 1918–
Bolivia

Artist-craftsman; family of artists. Studied painting and sculpture, jewelry and silversmithing in La Paz. Toured Bolivia and Peru studying folk-art and also dancing. From 1940 travelled in USA and danced with Martha Graham and Erika Thimey in New York. Since 1946 worked mainly as goldsmith and jeweler. Work shown Scranton Museum, Pennsylvania, and various galleries New York; Mexico 1954, La Paz 1956, São Paulo, Madrid, Rio de Janeiro, Buenos Aires and Berlin 1959, Bogota 1961. Lives La Paz.

Price, Arthur & Company Limited
Britain, Birmingham

Cutlers and silversmiths, founded 1902. 1966 completed new rolling mill and stamping shop at Parkfield, costing £100,000, designed to produce spoons and forks at rate of 2000 gross per week, that is, a finished spoon every five seconds, 80 per cent stainless steel, 20 per cent EPNS. Head of firm John Price. Have commissioned designs from Kenneth Lessons, Neil Harding, and David Carter, sponsor annual travel award at Birmingham College of Art. Subsidiary companies: John Mason (Sheffield) Ltd, Purcell Bros. Ltd, Parkfield Rolling Mills. Main brands: Atlas stainless steel, Arden plate. Over 400 employees in Group 1967. See pages 75–7

Puiforcat, Jean 1897–1945
France

Silversmith. Learned in workshop of his father, L. V. Puiforcat, who had one of finest collections of old silver in existence. Also studied at Central School of Arts and Crafts London. Most active silversmith in France in the 1920s; died on his return from Mexico.
Chevalier Légion d'Honneur, Croix de Guerre 1914–18, Rapporteur of the Paris exhibitions, 1925, 1937. Founder member 'Union des artistes modernes', silversmithing in Luxembourg and Toledo cathedrals and Solesmes Abbey. Sculpture: monuments to Descartes, Maison de France, La Haye; to Yves du Manoir, Olympic Stadium, Paris.
Puiforcat's friend Raymond Templier called him 'the creator of modern silver' and quotes in relation to him Fénelon's saying: *'Tout ce qui n'est qu'ornement, retranchez-le.'* Plates 330–3

Rajalin, Börje 1933–
Finland

Studied School of Industrial Art, Helsinki 1955. Worked for Bertel Gardberg 1957. International Design Award, American Institute of Decorators 1960. Gold Medal Triennale 1960 for famous wall decoration screen made of silver circles and spokes, exhibited extensively with the Finlandia exhibition, Milan, Zürich, Amsterdam, London, Vienna, Madrid. Plate 138

Ramsden, Omar (baptised Omer, perhaps after Homer) 1873–1939
Britain

Father greengrocer who later dabbled in design and set up as artist-craftsman; mother from family of ivory cutters and dealers. Omar studied Sheffield as silversmith. Made mace at Royal College of Art, toured Europe for year, joined father as apprentice, met

Alwyn Carr, who was first sleeping partner then collaborator with dilettante aristocratic approach; and registered mark at Goldsmiths' Hall with him 1898, Ramsden designing and making the silver and Carr enamelling. William Maggs, cretonne designer, became his draughtsman, smartening Omar's very rough sketches. Modeller and manager was Robert Hewlett. Unlike Fabergé or Boulton, Omar did not make known the names of his craftsmen. 1914 Carr, who managed finances, joined Artists' Rifles. Their second workshop in Seymour Walk off Fulham Road, called St Dunstans, held 14 workers and apprentices, including L. W. Burt, L. G. Durbin, R. A. Massey. 1918 Ramsden registered his own mark. Designed wrought iron gates for Old Bailey, using four Belgian blacksmiths. Once working on plate for three cathedrals at same time – Bermuda, Colombo and Coventry. Had distinctive and most pleasant style, reminiscent of early tudor work. Made extraordinary number of ceremonial presentation pieces, including mazer bowls for city companies, corporations and firms; and very many crosses, chalices, etc. for cathedrals and churches. Designed the Master Mariners' chain of office. Carr remained a bachelor. Omar married widow Annie Emily Berriffe 1927 but had no children. The nurse of Emily's two children by her first marriage, Jeanne Etais, became secretary and enameller for Omar. Ramsden was Liveryman of Worshipful Company of Goldsmiths, founder member of Art Workers' Guild, and Chairman of Church Crafts League. Used to cycle to work. Plates 314–16, page 137

Ravinet d'Enfert
France

Founded Paris 1845 by E. Tonnelier. At first limited to gilding and silvering works, it was transmitted to A. Lejeune who started manufacturing as goldsmith. When taken over by Louis Ravinet, who increased production abilities, buying from S. M. Henry and opening factory for manufacture of forks and spoons at Mouroux, the firm never ceased to thrive. 1891 Charles d'Enfert joined Louis Ravinet in partnership and firm became Ravinet d'Enfert, as it has remained. Production of forks, spoons and silversmith items increased rapidly. 1923 sons André Ravinet and Jacques d'Enfert took over management. The firm has continued to expand and today produce very wide range of articles. Won gold medals at exhibitions in Lyon 1894, Brussels 1897, and International Exhibition 1937 at Paris.

Redfern, Keith 1935–
Britain

Silversmith. Studied Sheffield and Royal College of Art London; teaches Hornsey College of Art. Consultant designer at Elkington & Co Ltd, 1963–5. Staffed Worshipful Company of Goldsmiths' display at 1962 British Trade Fair, Stockholm, and took the Company's exhibition to T. Eaton store Toronto and Winnipeg 1965. Plate 395

Reed & Barton
USA

Founded in Taunton, Mass. 1824. Now produces wide range of sterling and plated flatware and holloware. Also large supplier to hotels, restaurants, airlines, etc. throughout US. In 1966, one of 23 exhibitors chosen to participate in 'American Showcase', a travelling show illustrating the history of consumer goods touring in many shopping centres. 915 employees. Management and ownership always in same family. Wife of present president, Roger Hallowell, is founder Reed's great-granddaughter. Subsidiary Webster Company of Attleboro, Mass. make sterling children's goods, picture frames and pins (120 employees); Eureka of Taunton, bought 1966, make containers of wood and flannel. See *The Whitesmiths of Taunton*. Plates 108 110 114–20 432

Reiling, Reinhold 1922–
Germany

Artist-craftsman. 1936–40 studied engraving; 1943–5 studied goldsmithing in Dresden; 1948–53 designer in industry. Since 1953 has been teaching in Pforzheim; own workshop there. Participated in several exhibitions in Germany, more recently world-wide. Now artistic manager for new jewels, graphic designs and exhibitions in the Reuchlinhaus Museum Pforzheim, where he also has a semi-public workshop. See *Goldschmiedezeitung* 3/1967

Riegel, Ernst 1871–1946
Germany

Goldsmith. 1890–5 studied engraving with Kempten. 1895–1900 assistant of Fritz von Miller Munich. Worked Munich till 1906, then Darmstadt. Professor from 1912 at Städtische Werkschule Cologne. Plates 293–4, page 134

Riemerschmid, Richard 1868–1957
Germany

Architect and designer. Studied painting at Munich Academy. 1897 founder member of Munich vereinigte Werkstätte für Kunst in Handwerk, with Pankok, Obrist and Bruno Paul; collaborated with these artists in designing the Salle Riemerschmid at Paris Exposition of 1900. 1901 interior design for Munich Schauspielhaus. Designs for furniture and silver flatware are typical of his distinctively simple and functional style. 1902–5 taught at Nuremberg Art School. 1912–24 Director of School of Applied Arts in Munich. 1926 Director of Cologne Werkschule. Plates 285 289, page 134

Rodgers, Joseph & Sons Limited
Britain, Sheffield

Cutlers. With registered trade marks dating from 1682, Joseph Rodgers & Sons Ltd have protected their reputation for high quality by acquiring businesses of all cutlery firms in Great Britain bearing names 'Rodgers' or 'Rogers'. The showrooms built about 1860 were frequented by visitors, among them royalty, from all parts of the world. 1911 published figures showed 36 of their workmen had completed 50 years' service with firm. 1854 stone was fixed on wall in their Norfolk Street factory, commemorating

Joseph Whittington grinder who by correct taste united by rare skill as a workman enhanced the fame of Sheffield in its staple manufactures. Numerous specimens of his workmanship are in the showrooms of Joseph Rodgers & Sons. His private worth equalled his skill as an artisan and this tablet has been erected to his memory by his fellow workmen in token of their estimation of his character and abilities, 1854.

1943 this stone was set up against a wall in the factory yard at Pond Hill. No one knew exactly what instrument he used as a grinder, and to this day the finely cut lettering shows no sign of decay.

J. R. Ackerley in *Hindoo Holiday* wrote:

As Babaji Rao and I were walking in the outskirts of the village this evening, two old peasants, a man and a woman, begged of me, the old woman was ill it seemed; she squatted on the ground at my feet and moaned and rocked herself, holding out her clawlike hands, while the old man, who was thin and hairy and almost entirely naked, begged for medecine for her, 'good medecine', he kept saying, 'Rogers medecine'.

Babaji was very amused, and explained to me that there were some steel articles of recognized excellence, marked 'Rogers' being sold at the fair, and the old man wanted some medecine as good as this steel.

'Of course he thinks you are a doctor', he said; 'these poor people think that all white men are doctors'.

I asked him to explain that I was unfortunately not a doctor, and I gave the old man a rupee to buy some 'Rogers Medecine' . . .

1869, Joseph Rodgers & Sons were producing each week 3000 dozen pocket knives, 80 dozen razors and 600 scissors. The Bowie knife, for example, was made for many years from the early part of the century by craftsmen of Joseph Rodgers & Sons, from the best materials.

The works in 1878 used 26 tons of ivory for handles and scales, comprising 2561 tusks, averaging over 22 lb each. About 1911 the storeroom held 15 tons of ivory that was valued at some £22,000. Several of the tusks weighed about 160 lb each, while a special tusk weighed 216 lb. Others included baby teeth, weighing only 2 or 3 lb. Today Rodgers are probably best known quality firm in Sheffield cutlery, as well as being the oldest cutlery firm in the world.

Rohde, Johan 1856–1935, Denmark. Painter, silver and jewelry designer. See page 11, plates 1–7 9 10 307

Rose, S. J. & Sons (Goldsmiths) Limited
Britain

Leading box-makers, founded 1915 at 21 Wardour Street by Stephen John Rose, now managed by Julian Stephen Rose and Arthur Brian Rose. Acquired in London Watts & Rumball medallists, Louis Auvail, Mappin & Webb box factory; Cohen & Charles, and H. Huntley, Albert Carter, F. Field of Birmingham. Main workshop now 19 Ridgmount Street, London, producing a wide range of gold, silver and enamels.

Saglier Frères
France

Founded Paris 1842 by Victor Saglier (1809–94). A family business, grandsons of founder being present owners. Started in small way by importing and selling English silverware. As business increased they were obliged to open small workshop for assembling, repairing and electro-silver plating. This was later to become main undertaking of firm. On the death of their father in 1894, Eugène (1859–1935) and André (1862–1948) became partners. This proved very successful alliance owing to good business sense of the one and technical knowledge and artistic dispositions of the other. They expanded rapidly. André Saglier was responsible for creating most models of flatware and holloware which brought firm many flattering awards at International Exhibitions from 1889 on. After André's death in 1948, his sons, Jean and François, took over firm, having been assistants for many years. The fourth generation of Sagliers are now preparing to assume management.

Schaffner, Alexander 1929–
Switzerland

Studied La Chaux de Fonds and Cologne. 1956 joined the Swiss Society of St Luke. 1961 joined SWB (Schweizer Werkbund). Designed cutlery for Pott of Solingen 1960.

Schlanbusch, Ellen 1910–
Denmark

Engraver and silversmith, designs for Just Andersen mostly in pewter; after his death in 1943 became manager of firm, also produces gold and silver jewels.

Schlumberger, Jean 1907– . See pages 47 53, plates 94 96

Schnellenbühel, Gertraud von 1878–
Germany

Painter and designer. First studied painting in Munich. Turned to applied art 1902 and attended classes in metalware at Debschitz school. 1911 joined workshop of silversmith Adalbert Kinzinger. Represented at Deutscher Werkbund exhibition Cologne 1914. End paper

Shiner, Cyril James 1908–
Britain

Trained under Bernard Cuzner at Birmingham and taught at Bournville and Vittoria Street schools. He and R. J. Ruby went to Royal College of Art at same time, after which Shiner went back to Birmingham to teach, design and make. Designed notable ceremonial pieces. See page 137

Shreve, Crump and Low
USA

Retail jewelers and silversmiths established 1800 in Boston by Macfarlane family. The Shreve family joined management in 1854. Richard S. Shreve, present head of firm, is the fourth generation and his children may succeed him. One-man show of jewels by Andrew Grima of London, in association with the Boston Museum of Fine Art 1967. Group exhibition of British goldsmiths' work with Worshipful Company of Goldsmiths 1967.

Silver Workshop Limited
Britain

Founded by R. W. Stevens and Tony Laws, later with Keith Redfern. Stevens studied at Gravesend School of Art under Jack Stapley and at Royal College of Art. Teaches at Hornsey and at Central School of Arts and Crafts. Won Worshipful Company of Goldsmiths' competition for Everest Trophy 1953 to commemorate the first ascent of Mount Everest by the British team. Registered mark at Goldsmiths' Hall 1964. Plates 394–5

Silver, R.
Britain

Design studio at Corner House, Brook Green, Hammersmith, London. Founded 1880 by Arthur Silver (d 1896) and continued by son Rex (1879–1965), who was later joined for a few years by brother Harry. The Silver studio designed for decorative arts: textiles, wallpapers, floor coverings. 1899 Rex Silver turned to silver design, being one of first to work for Liberty's in their 'Cymric' range of silver and jewelry. A pair of Cymric silver candlesticks and a pewter bowl designed by him are in Victoria and Albert Museum. Contributed to many exhibitions including Paris 1925 and 1937 and the exhibition of Historic and British Wallpapers 1948. Rex Silver was Fellow of the Royal Society of Arts and one of first people to be appointed to National Register of Designers.

Slutzky, Naum J. 1896–1966
Britain

Born in Ukraine. Educated and apprenticed in Vienna: study of fine arts, then of engineering at the Vienna Polytechnic. Worked at Wiener Werkstätte in goldsmiths' department. Later freelance designer. Appointed master to department of product design, Bauhaus, Germany. After Bauhaus, freelance industrial designer in Hamburg, until 1933, when emigrated to England. Tutor of metalwork and engineering at Dartington Hall until war; research on production of diamond-tools used in optics, 1946–50, part-time tutor at Central School of Arts and Crafts, London; 1950–7,

established and equipped product design section of School of Industrial Design at Royal College of Art. 1957–64, head of School of Industrial Design at College of Art and Crafts, Birmingham. Jewels shown at Goldsmiths' Hall, Design Centre, London 1964. Freelance designer. The 1965 head of industrial design department, Ravensbourne College of Art Bromley. An expert enameller, he tried to use repeated units in different combinations to give character to his jewels, but clients preferred handwork and he never mass-produced jewels as he would have wished, because of his continuously active and original teaching. Naum's parents fled the pogroms in Kiev, and he remembered as a child hearing Tchaikovsky conduct in Vienna.

Spencer, Edward 1872–1938
Britain

Founded Artificers Guild 1906, headquarters in Conduit Street and branch in Kings Parade, Cambridge. Employed orphan boys as apprentices. At height of his success had staff of 40. Superb draughtsman, combining the use of different materials in association with silver: gourds, nuts, ivory, shagreen, mother of pearl and wood were worked into his designs.

Spink
Britain

Leading dealers in classical antiquities, silver and jewels, coins and medals. John Spink, son of yeoman farmer at Kirkthorpe, near Wakefield, Yorks, apprentice goldsmith 1657–66. Set up as goldsmith Lombard Street 1666, joined 1670 by cousin Elmes Spink, pawnbrokers, dealers. Shopwindow necessary, so in 1703 moved to Gracechurch Street, where they stayed over 200 years. David and Philip Spink now directors, also Anthony in firm. Celebrated tercentenary October 1966 with important display in King Street showroom, 4000 years of works of art. Medal factory in Raynes Park. See *Connoisseur,* Dec. 1960

Stabler, Harold 1872–1945
Britain

In charge of art section of Cass Institute. Partner in firm of Poole Potters. Some of tiles he designed for Pick and London Transport can still be seen in tube station at St Paul's and elsewhere. Designed many presentation pieces for Goldsmiths' and Silversmiths' Company and Wakely & Wheeler. Firth Vickers, for whom Brearley had invented stainless steel, commissioned Stabler to design series of prototypes including original etched techniques, but no British producer was interested; stainless steel was therefore exploited abroad, to Britain's loss. See pages 88 137, plates 144–5 253 318 324

Stapley, Jack E. 1925–
Britain

One of most successful post-war silversmith designers and craftsmen. After leaving Royal College of Art taught at Gravesend school of art, where he built up thriving department of silversmithing. Later became instructor at Central School. In his workshop, which he fitted up himself, he made a number of maces and other ceremonial pieces of intricate design. Won 1951 competition of Worshipful Company of Goldsmiths for rosebowl to celebrate Festival of Britain.

Steltman, J.
Holland

Noordeinde 42a, The Hague. Retail and manufacturing jewelers. Founded 1917 by late Johan Steltman (1891–1961) who studied Hanau. 1925 Grand Prix Paris exhibition for a teaset. Mr Jan R. Bakker now principal. Diamond watch designed by Piet van Belkum given to HM the Queen on her 50th birthday by Holland.

Stephensen, Magnus 1903–

See Jensen, pages 10 12, plates 20–1 342

Steward, William Augustus 1867–1941
Britain

Head of silver section of Central School of Arts and Crafts in the 1920s. A flamboyant, colourful character, he occasionally rode horses into the school. Editor of *Watchmaker and Jeweller* and *Gemmologist*, two of the trade papers; on staff 1895–1934.

Stone, Robert E. 1903–
Britain

After apprenticeship, held Worshipful Company of Goldsmiths' scholarships, and set up workshop at 22 Garrick Street. Teaches; among his pupils his daughter, Jean, a jeweler.

Styles, Alex G. 1922–
Britain

On permanent staff, as designer, at Garrards since 1947, during which time he must have designed as much silver as any other person alive. See Garrards. Plate 378

Sunyer, Ramon (Ramon Sunyer Clara) 1889–
Spain

Jeweler and silversmith. Apprenticed to father (also a silversmith) at 13; family workshop dating back to 1835. First jewelry shown Sala Pares 1912. Also exhibited Paris 1925 (Gold Medal), Barcelona 1929 (Grand Prix and Gold Medal), Milan Triennale 1936. 1955 retrospective exhibition Barcelona; also Grand Prix d'Honneur for jewelry and enamels, III Hispano-American Biennale.

Swaffield-Brown, Thomas 1845–1914
Britain

With Hunt & Roskell until 1887, winning National Medallion at age of 16. 1887–1914 took charge of art department of William Hutton & Sons, as head designer. Particular success with modelling human figure, and small equestrian statues. Took an active part in founding Sheffield Art Crafts Guild and in introducing silversmiths' classes at Technical School of Art. An authority on stained glass, examples of his work in churches. Also a talented musician, especially in church music; he was chairman of the local board of Trinity College of Music, London. Latterly Huttons paid him the high salary of £1000 annually, and his son George Brown who worked with him, £500 or £600. Brown was a figure typical of the devotion and convention of his day.

Thomas, David 1938–
Britain

Jeweler and silversmith. Studied 1955–7 Twickenham School of Art, 1959–61 Royal College of Art. Own workshop 1960. 1958–9 worked Stockholm under Sven Arne Gillgren. Since 1963 has taught Central School of Arts and Crafts London. Helped staff exhibitions by Worshipful Company of Goldsmiths, Zürich 1963, Sydney 1964. Designed mace given by Worshipful Company to Newcastle University. 1967 one-man show at Famous Bar Store, St Louis, Missouri, USA.

Tiffany, Louis Comfort 1848–1933

See pages 49–53; Mario Amayo in *Apollo*. February 1965; New York Museum of Contemporary Crafts catalogue, 1958. Plates 94–6 104–6

Tillander
Finland, Helsinki

Royal Swedish Court Jewelers. Established 1860 by Alexander Tillander, a Finn, in St Petersburg when Finland was part of the Duchy of Russia. His jewelry shop prospered and he became a supplier of jewels to Russian Court. When Revolution broke out Alexander Tillander was killed by his own workmen but his son, grabbing a handful of diamonds, escaped on last train out of St Petersburg back to Finland. Firm now well established in Helsinki, with branch in Stockholm. They retail silverware at 48 Alexanderst. but specialize still in custom-made jewelry from their factory on the outskirts of Helsinki and in medals and decorations. Oskar Pihl, once Fabergés chief workmaster designate, now their chief designer, designed the medals for 1952 Olympic Games. Present President of firm is Herbert Tillander, founder's grandson. He is a gemmologist of world renown, being a member of the Gemmological Associations in Britain, America, Sweden and Finland. Plate 126. See page 23.

Torun (Torun Bülow Hübe) 1927–
Sweden

Artist-craftsman. Studied Stockholm. Own workshop

1951. Moved Paris 1956, works Paris, Biot. Silver medal Milan Triennale 1954, Gold Medal 1960, Lunning Prize 1961. Worked for Christofle Paris 1951, Orrefors Sweden 1951–61. Shows regularly at Galerie du Siècle, Paris. One-man show Jensen, New York 1963.

Tostrup, Jacob
Norway

Karl Johansgt 25, Oslo. Manufacturing and retail firm, founded 1832 by the last Worshipful Master of former Oslo Goldsmiths' Guild: Jacob Ubrich Holfeldt Tostrup, great-grandfather of present manager. In last half of 19th century revived country filigree work and exported it all over the world, soon being followed by others including David Andersen. Oluf Tostrup was influential in founding Oslo Museum and Royal School of Design. In 1890s produced remarkable *plique à jour* enamels which enjoyed world-wide exhibition successes, particularly the cases and cups designed by architect Torolf Prytz *c.* 1900, when he was head of firm, having married founder's grand-daughter Hilda Tostrup. Torolf was president of exhibitors for Norwegian centenary celebrations 1914. Jacob Prytz, now firm's proprietor, as head of firm after 1918, and teacher, later principal at State Handicrafts and Art School, worked with Thor B. Kielland, curator and later director of Oslo Museum of Applied Art, to introduce new design into Norway. Through Society of Applied Art, Foreningen Brukskunst, founded 1918 (widened after 1945 under name National Association of Norwegian Applied Art – Lands-forbundet Norsk Brukskunst), they organized notable 1920 exhibition in garden suburb of Ullevål. For 1928 Norwegian Exhibition at Bergen, they persuaded some factories for first time to produce modern work, for instance Hadelands glass with the designer Sverre Pettersen, or Porsgrunn porcelain with the young Nora Gulbrandsen, or stoves from the Jøtül foundry, with the architects Gudolf Blakstad and Herman Munthe-Kaas. In 1930s under Prytz, the architect Arne Korsmo (his son-in-law) and later Arne E. Holm, helped to form the school's progressive policy. Torold Prytz, Jacob's son, is now the executive.

1954 Arne Korsmo's silver-plated cutlery, and in 1952–4 Mrs Grete Prytz Korsmo, Jacob Prytz's daughter's enamelled silver bracelet and ring, all for Tostrup, showed the firm giving mechanized production a distinctive art form. At 1958 Brussels World Fair Tostrup exhibited spectacular enamels on large stainless steel bowls in Norwegian government pavilion. Norwegian water has qualities said to facilitate enamel; be that as it may, Tostrup have made Norway the recognized home of modern enamelwork, winning for it a Grand Prix in Paris 1900, and 1937, and at the 10th Milan Triennale 1954. Mrs Korsmo won the Lunning prize in 1953, and a Grand Prix at the 1954 Milan Triennale for enamel, while her husband, architect Arne Korsmo, won another Grand Prix for the whole Norwegian display. Plate 303. See page 334, Korsmo, Eriksen

Towle
USA

Founded 1690 Newburyport, Mass. by William Moulton. Business continued through six generations of Moultons and in 1857 it was passed on to Anthony F. Towle and William P. Jones, two Moulton apprentices. 1882 firm became Towle Manufacturing Company. Towle makes only sterling silver articles. Has wide range of flatware and holloware. Silver by Moultons is found today in museums and private collections. Examples of their work are a part of Towle collection of early American silver exhibited in Towle Gallery, Newburyport. They frequently sponsor distinguished exhibitions such as 'Sculpture in Silver from Islands in Time' (an American Federation of Arts exhibition), 'The Odd and the Elegant in Silver' (from museums and private collections), 'Knife/Fork/Spoon' (Walker Art Center, Minneapolis).

Treskow, Elisabeth 1898–
Germany

Goldsmith, jeweler. Studied Hagen, Essen, Schwäbisch Gmünd; also in Munich under Karl Rothmüller. Church furniture as well as jewelry. Workshop in Essen 1919–43; Detmold 1943–8; Cologne since 1948, where she is head of metalwork at the art school. Exhibited widely. Gold medal Milan Triennale 1936; Paris 1937; Grand Prix Triennale 1940. Research into gold granulation. Designed cutlery for Pott of Solingen. Plates 65 177, pages 38 138

Tyssen, Keith 1934–
Britain

Studied Sheffield College of Art and Royal College of Art London. Own workshop Sheffield 1963 making special commissions, teaches Sheffield. Made some silver for Guildford Cathedral and 1967 for Churchill College Cambridge chapel, both given by the Worshipful Company of Goldsmiths. Plates 380–2

Van Cleef & Arpels
France

Court Jewelers Paris; New York; London; Palm Beach; Geneva; Cannes; Monte Carlo; Deauville. Founded 78 years ago. Invented the minaudière, a word now commonly used to describe a type of box; and the invisible setting – a type of pavé setting in which the metal holding the square in position cannot be seen. The Arpels family are now the firm's principals.

Van de Velde, Henri Clemens 1863–1957
See pages 7 30 56 72 135, plates 271–81

Van Kempen & Begeer
Holland

Founded 1835. Now largest producers in Holland with factory at Zeist, workshops at Voorschoten and Coevorden, retail shops in Amsterdam and other cities, and agencies throughout the world. Head of firm is Sebastian Begeer who, like his father, is a keen supporter of BIBOA (Bureau International de Bijouterie, Orfèvrerie et Argenterie). Art Director Gustav Beran (b 1912 Vienna), student of Josef Hoffmann and Eugen Mayer. A special department has now been opened for hand-made work with Gijo Bakker and Frans Brusche (b 1902); their metal, wood and enamel wall-relief plaques made a distinguished showing at the 'Internationales Kunsthandwerk' Stuttgart LGA Museum show 1966. The main Dutch centre for silver is Schoonhoven, but the factories there are relatively small compared with Begeer. See Beran

Vander, C. J. Limited
Britain

1886 C. J. Vander bought Macrae & Goldstein, Covent Garden, later joined by sons Henry and Alfred Vander. 1907 moved to Betterton Street where founder's three grandsons Arthur, Henry and Norman joined; 1938 moved to Fetter Lane. 1948 acquired Francis Higgins cutlers from the old Garrards. 1958 the flatware business of Atkin Bros Sheffield, one of the most famous firms whose works name Truro Works referred to their trade in Britannia metal using Cornish tin. Atkins had themselves absorbed the fine old company Thomas Bradbury in 1947, whose name lives on in Frederick Bradbury's *Old Sheffield Plate*. 1965 Vanders bought Roberts & Belk Sheffield, and Benton Bros one of the only remaining firms of casters. H. M. Lister, manager of Roberts & Belk, joined them 1905, retired 1966, typical of the long and loyal service in the older Sheffield factories. 1967 opened new Sheffield factory for Roberts and Belk's 150 people, including William Bush with whom they had been associated for 30 years. Plates 144–5

Viners of Sheffield
Britain

Broomhall Street, Sheffield. Manufacturers. Headed by chairman and managing director, Ruben Viner, Viners is British silver trade's biggest employer with payroll of 1100 out of the Sheffield total of some 11,000 workers spread around 225 firms mostly with only five employees apiece. Recently installed £15,000 triple-acting hydraulic press, one of first ever used for silver holloware; press turns out shells of deep drawn coffee and tea pots in one draught, instead of six when sheet silver is banged out by hand. Started by Ruben Viner's father and two uncles in 1907 in Sheffield, old home of silver trade: his sons, Brian, now in charge of manufacturing, with Roger on sales. Leslie Glatman sales director. Gerald Benney retained as designer. His 'Chelsea' pattern with alternating polished and matt surfaces was probably first British stainless cutlery entirely machine-made, the result of semi-flow line production which took four years to evolve, 1958–62; almost untouched by hand, it is a world best seller, as is also his 'Studio' stainless steel, with the world's first abstract ornament in high relief. In 1967 'Studio' sold

some 1500 dozen pieces weekly, 'Chelsea' 3000 or 4000 dozen pieces weekly. These enormous sales are achieved by excellent design suited to production efficiency. But retailers are interested also because Viners allow them an exceptionally high profit margin of as much as 100 per cent. Seven pieces of 'Chelsea' in 1967 cost £2 9s 6d retail; a similar place setting of 'Studio' £3 3s, while the new 'Design 70' was £2 12s 6d. 1967 Benney's third highly mechanized cutlery scheme 'Design 70' introduced. Its shape is controlled by need for easy machine polishing, by the wish to give biggest possible weight to this light material, and by wasting as little as possible, making blank shapes interlocking and symmetrical, and by conscious effort to be English, echoing 18th-century fiddle pattern with its considerable length and weight.

1964 Lord Queensberry designed economical stainless steel holloware with teapot lid hinging in front. 1960 Viners cutlery was 85 per cent EPNS, in 1967, 75 per cent stainless steel, including pieces made in Viners Hong Kong factory and in Japan. 1967 the total flatware weekly production was about 25,000 dozen, or nearly half a million, pieces, and an order for three million pieces for contract use is quite normal. Viners have about 1500 holloware patterns in production and they make about one million objects each year. Their turnover has increased by 4 or 5 times since 1945; they bought two fine old Sheffield firms of cutlers, Thos. Turner, and Harrison Bros. and Howson. They have always concentrated on lightweight sterling silver tableware of traditional design, of which they now make some 60 per cent of entire British production. They also have big output of plated wares, often with elaborate floral ornament resembling chasing but in fact pressed on by embossed steel rollers. World-wide exports: 1967 opened first German warehouse and offices at Gereonstrasse 18–32, Cologne.

1967 Viners organized important international stainless steel tableware competition with four British and two foreign judges. There were 670 entries from 33 countries but no awards were made for holloware. A student at Hornsey College of Art, Robert Glover, won the first prize of £1000 for his economical and dignified cutlery, chosen from 250 designs from 27 countries, which may now, therefore, be put into production. The contest has thus begun to produce fame and success for an unknown young man, a gratifying result. Viners latest cutlery enterprise is, however, with Benney, the no scrap blank, by which expensive scrap is almost totally eliminated, a major breakthrough. There will be no waste, no unwanted metal to be disposed of – and there is no use for stainless steel refuse. Such is real industrial progress, a continuous quest for the ideal mechanical cycle of production. Plates 189 427, pages 75 138 140 142

Wagner, Otto 1841–1918
Austria

Architect, professor Vienna Academy 1894. Leading

designer in Sezession style which he abandoned for a purer functionalism about 1904. His suburban stations in Vienna are famous. Plates 257–60

Wakely & Wheeler Limited
Britain

Silversmiths. First maker's mark registered in 1791 by John Lias; then 1818 John Lias and Henry Lias; 1823 John Lias, Henry Lias and Charles Lias; 1837 John Lias and Henry Lias; 1850 Henry John Lias, Prime Warden of the Worshipful Company of Goldsmiths 1861, and his son; 1879 Henry John Lias and James Wakely; 1884 James Wakely and Frank Clarke Wheeler; 1909 Wakely & Wheeler. Arthur Day Wakely was Prime Warden of Goldsmiths' Company 1939. Always had fine craftsmen, including W. E. King and F. S. Beck, who taught at the Hornsey School of Art and at Central School. After Arthur Wakely's death firm was taken over by Padgett and Braham Ltd, box and insignia makers under Stanley Padgett, who also acquired R. E. Stone's business 1964, Pairpoint's gilders 1964, Manton's box-makers Birmingham 1965. A. E. Pittman, Wakely's partner, did much of designing. Kenneth Mosley was for many years designer in their factory in Red Lion Square, London, before the firm moved to Ganton Street. Plates 322 324 369 374 377–9 393 396–7

Walker & Hall Ltd
Britain

Sheffield silversmiths. Founded 1843 by Hall, a lawyer and man of affairs, in partnership with Walker, an artisan, to exploit new invention of electroplate. The firm used to claim that George Walker plated the first 'useful articles' in Sheffield, but real credit belongs to Elkingtons, with their early licensees Harrisons and Huttons. This was no more than a publicity effort by Sir John Bingham, Hall's nephew, who joined the firm at 16 and had brought it to its peak by his death in 1916. It was the top firm in the trade. Its main business was contract; the general public obtained personal introductions and bought at wholesale rates.
From 1916 under Sir John's son Sir Albert Bingham, the steady expansion stopped. Sir Albert devoted only part of his time to the firm. He died in 1945 and in 1954 Peter Inchbald, who scored a great victory in Worshipful Company of Goldsmiths' open national competition for a new mace for the University of Hull in 1955, and who was a great-grandson of founder, and his brother-in-law Lt-Col. J. T. A. Wilson joined the board, its only members under 60. A big modernization programme ensued with advice from Urwick Orr, management consultants. With David Mellor as consultant designer, Peter Inchbald became the one champion of modern design in the British industry. He served on Council of Industrial Design and the firm, which had never before shown artistic initiative, won several Design Centre Awards. Mellor's 'Pride' pattern cutlery which he had made at Royal College of Art and been trying to sell to a manufacturer for three years, came first with a Design Centre Award in 1957, and by 1963 represented 30 per cent of the firm's EPNS flatware output! Then there was the teaset, which won a 1959 Design Centre Award but was less of a commercial success. The 'Spring' or 'Campden' flatware by Mellor and Welch followed, a joint venture with J. & J. Wiggin in stainless steel – Walker & Hall's first adventure with what they persisted in regarding as a new material, though by then it was 25 years old. 'Spring' is now a leading stainless steel pattern of the giant British Silverware group. Other fruits of Walker & Hall/Mellor combination were 'Symbol' flatware (1962 Design Centre Award) and 'Fanfare' tea and coffee sets for catering, and the newest, 'Minim' for government canteens. Sadly the firm's antiquated system of giving large discounts through its many semi-private retail offices, always unattractive to the normal retail trade, proved too complex to modernize in time to absorb new production; the new production techniques themselves were perhaps too sophisticated – as Peter Inchbald himself says, 'we moved from bows and arrows to guided missiles, when what we really needed was an honest to goodness gun.'
Since 1963 when British Silverware bought the company, the original Walker & Hall factory with its commanding site by Sheffield City Hall, surmounted by the huge W. & H. trade mark pierced on an iron triangular pennant, has been demolished. The factory established in a partly new building at Bolsover was closed, became a warehouse, and its machinery moved to Elkingtons in Walsall. But the daring modern impetus was not entirely lost. Seven smart new retail shops have been opened first in Liverpool then in Manchester, Sheffield, Edinburgh, Newcastle, Bristol and Bournemouth, the architects being Forrest and Reid, under the management of John Thomas. Eric Clements is now in charge of the group's modern design. Plates 385 424, pages 75–6 139

Wallace
USA

Founded 1835 Wallingford, Conn. by Robert Wallace, to make German silver spoons. 1871 became R. Wallace & Sons Mfg Co. Now produces all types of flatware and holloware, both sterling and plate.

Warham Guild
Britain

Church furnishers, named after William Warham, the last pre-Reformation Archbishop of Canterbury; founded 1912 by Percy Dearmer, Vicar of St Mary's Primrose Hill, later Professor of Ecclesiastical Art at King's College, London. Compiler in 1906 of the 'English Hymnal', connected with the foundation after 1963 of the Council for the Care of Churches.

The Guild's trading profits, with an issued capital of 4000 £1 shares, cover its liabilities, and it tries to pay dividends. It handles commissions and sales from its London showroom, 28 Margaret Street. Its advisory committee normally includes, as well as ecclesiologists, several bishops and deans. See page 131

Wartski, C. & H. Ltd
Britain

138 Regent Street, London and in Llandudno. Retail jewelers and silversmiths. Founded Llandudno 1865; 1962 London shop redesigned by Denys Lasdun. The firm specializes in the work of Fabergé, which was first popularized after Bolshevik Revolution by Emanuel Snowman, now aged 81. He visited Russia fourteen times, almost annually after 1926 his first visit, buying these aristocratic treasures from Russian government in exchange for much needed foreign currency. The government also auctioned some of Russian Crown Jewels at Christies in 1927. Emanuel opened in London after marrying the daughter of Maurice Wartski, the Llandudno owner. Emanuel has now been twice Mayor of Hampstead. Kenneth Snowman is now executive of the firm and with his books *The art of Carl Fabergé* and *18th-century gold boxes of Europe* has made a fine contribution to the study of gold and enamel. Plates 87–94

Weckström, Björn 1935–
Finland

Studied Goldsmiths' School, Helsinki 1956. Travelled and exhibited extensively. Second and third prizes, Hackman cutlery competition 1959. Won 1965 world jewelry competition organized by Stern of Rio de Janeiro. One-man show 1966 at Nordiska Company store, Stockholm. 1957 designer at Hopeakontu Oy; own workshop 1957; designer at Kruunu Koru (Crown Jewelry) since 1964.

Weiss, Kurt 1911–
Britain

Born in Vienna, came to England 1938 and set up small workshop in the Barbican in the City of London, manufacturing gold boxes; employed one man. Weiss's wife worked as polisher. Interned for nine months at beginning of war. After war, due to crippling purchase tax nearly all his work went to Far East, a market since closed due to embargo on gold imports in India, etc. Bombed out of Barbican, set up at present address, Carthusian Street, and considerably expanded. Now perhaps the best box-makers in Europe. His son Tony, after two years' study of silversmithing and diamond mounting at Sir John Cass College, has recently joined the firm. See page 76

Welch, Robert 1929–
Britain

Artist-craftsman and industrial designer; studied Mal-vern, then Royal College of Art. At Chipping Campden, Gloucestershire, since 1955 in workshop building originally converted by C. R. Ashbee's Guild of Handicraft. Consultant designer amongst others to J. & J. Wiggin of Walsall, pioneers of stainless steel holloware production in England since the 1930s, who now produce his stainless steel tableware named after liner *Oriana*. Silver medal, Milan Triennale 1962. Has also made jewels and designed a range in base metal for Messrs Dennmark 1961. See 'Welch' by Graham Hughes in *Connoisseur* 1963. 1967 range of enamelled steel kettles etc. designed for Carl Prinz, Solingen & Wagenfeld. One man show Heals, London, Design Centre Shop, Copenhagen. Plates 144 146–9 390–1 393, pages 56 90 138–40 142

Wellby, D. & J., Britain

18–20 Garrick Street, Covent Garden, London, WC2. Retail and manufacturing jewelers and silversmiths. Founded 1820 by John Wellby in King Street Soho, near present Saville theatre, as bullion dealers. He used to walk home with bags of gold to Highgate. In 1860s became bullion brokers and bankers; John found diamond dealing more profitable, however. Customers wanting to cash cheques often found no ready money because John had spent firm's assets buying diamonds. 1865 started a 99-year lease of present premises. 1896 became a limited company managed by John's two sons, Daniel and John. Sir Squire Bancroft and Sir Johnson Forbes-Robertson first commissioned silver from firm which is near the Garrick Club, a favourite for actors; but it was only after 1945 that gold melting equipment in Garrick Street was removed and bullion dealing discontinued in favour of jewelry and silversmithing business. Firm owns a receipt dated 26 March 1873 worded as follows: 'received of Mr. Wellby the sum of £8,410 for account of the Right Honble. B. Disraeli, M.P. for jewels. (signed) N. M. Rothschild'. Founder's grandson, Edwin Wellby, was Prime Warden of Worshipful Company of Goldsmiths in 1930, and present principal of firm, his son Guy Wellby, is now a past Prime Warden.

Whiles, Gerald 1935– , Britain

Studied Birmingham College of Art then Royal College of Art London. Won first prizes in Ascot Cup competitions – Royal Hunt Cup 1961. Designed and made for the Company: trophy for the best stand, New York Exhibition 1960 (given to Tube Investments Ltd). Helped to staff the Company's exhibition, Stoneleigh Abbey, 1959. Now freelance designer, teaches Bourneville College of Art. Made British Government's gift to Tanganyika at the time of her independence, also silver given to Guildford Cathedral by Worshipful Company of Goldsmiths. Plates 392 394

Widmer, Mrs M. 1928– , Switzerland

Artist-craftsman. 1944 Zürich art school; 1946 Lucerne art school; 1954–6 designed for Gubelin, then studied

Rome; since 1956 head designer for Gubelin. 1958, 1959, 1960 Diamonds-International Awards, New York. 1963 Prix de la Ville de Genève for an unusual clock, in a block of rough amethyst encrusted with gold, which is wound only once a year. Personal style of repeated circles and bars.

Wiggin, J. & J. Limited See page 56, plates 144–9

Wilm, H. J.
Germany, Hamburg

Established 1767 Berlin by Gottfried Ludewich Wilm. An excellent goldsmith and craftsman, Wilm attracted attention of King Frederic the Great of Prussia. Was joined in his workshop by his son Heinrich Ludwig. They built up silver workshop and Heinrich added department for jewels that became known as the 'Jewel Case'. Recognizing the qualities of Heinrich, the King appointed him royal silversmith and granted him Order of the Red Eagle. Firm passed to Heinrich's nephew, Hermann Julius Wilm (b 1812). He also was appointed by royal house to design plate for household and Treasury. In 1859 workshop moved and was modernized. Under Ferdinand Richard Wilm at the beginning of 20th century, sterling silver was introduced to German craftsmanship, and design revived in Berlin with such famous artists as Peter Behrens. 1905–7 Wilm helped to found Capetown–Cairo Railway and Telegraph Company. He spent five years in London 1906–11. Appreciated British sterling standard of 92·5 per cent purity, which he found much superior to normal continental and German work with its workshop for this purpose. 1911 appointed royal jeweler to courts of Bulgaria and Rumania; 1912 to the German Emperor and Empress. He campaigned for sterling silver through German trade association of which, in 1919, he joined the management, and through Deutscher Werkbund. In 1937 the firm won a gold medal for its sterling at Paris World Fair.
The 1939–45 war destroyed factory and business in Berlin and in 1948, under direction of Johann Renatus Wilm (sixth generation to manage the business), because of extremely difficult economic and political conditions the firm was moved to Hamburg. Now has interest in Idar Oberstein stone-cutting works.
1926 F. R. Wilm, already campaigning for sterling silver, widened his activities. With his friend Carel J. A. Begeer of the Dutch silversmiths, he founded BIBOA (Bureau International de Bijouterie Orfèvrerie et Argenterie) which held and still holds regular international conferences, which, however, during Nazi and post-war periods were not truly international because of natural political feelings. 1932 with Wilhelm Waetzoldt, Director of the Berlin museums, he founded Deutsche Gesellschaft für Goldschmiedekunst to help the goldsmiths' art with competitions, exhibitions, conferences and prizes. Despite lack of funds this society did very useful work in the 1960s in Hamburg under Ulla Stöver its secretary. 1932 he initiated society's scheme to give golden rings of honour (der goldene Ehrenring) to outstanding craftsmen, each ring being designed and made by a previous winner, and the awards being almost every year. Till 1966 there were 27 winners, 21 of them German. See page 73

Wilson, Henry 1864–1934
Britain

Trained as architect, joining J. D. Sedding's office as chief assistant, succeeding to his practice on Sedding's death in 1891. c. 1895 set up workshop for metalwork and jewelry. c. 1901 onwards taught metalwork at Royal College of Art. Author of *Silverwork and Jewellery*. A great cup surmounted by a figure of St George and the Dragon by Wilson was prominent in British Pavilion at the Paris exhibition of 1925. Worked at Torcello Venice at memorial tomb to Bishop Elphinstone of Aberdeen University. 1922 moved to Paris, worked on doors for Salada Tea Company of Boston, then on doors for New York St John the Divine Anglican cathedral; memorial pulpit in Ripon cathedral, memorial in St Mary's Nottingham to his wife's father Canon Morse. President Arts and Crafts Exhibition Society. Master Art Workers' Guild, increasingly active lecturer. See Cooper, Murphy. Plates 254–6, pages 133–6

Wirkkala, Tapio 1915–
Finland

Designer, etc. Studied sculpture Helsinki (Central School of Arts and Crafts) 1934–6. Designed many exhibitions all over the world. Seven Grands Prix Milan Triennale, more than any other designer in the world, the first 3 in 1951 for glass lights, woodcarvings and design of the Finnish Section; 3 in 1954, and one in 1960 for electric bulbs, with a gold medal. Chief designer glass factory Karhula-Littala from 1946. Art Director of Central School, Helsinki, from 1951–4. Lunning Prize 1951. Designed first jewels for Goldsmiths' Hall, London 1961 exhibition, and, for Christmas 1966, a series of magnificent gold pendant jewel masks, after seeing Mycenean masks in the Athens Museum. Honorary Royal Designer of Royal Society of Arts in London 1965. Designs for many factories. His wife Rut Bryk designs for Arabia Pottery. Plates 365 438–41

Wise, Theodore C. F. 1898–
Britain

Engraver. Trained at Central School of Arts and Crafts in London where he won a scholarship; later taught there. Founded firm T. and A. Wise of Broadwick Street, London, with his brother Mr A. de G. Wise in 1945, with specific intention of improving standard of engraving in London, and keeping it at highest level possible. He is now perhaps the world's leading engraver. He usually employs between 8 and 12 craftsmen. Firm has won many prizes for its superb work. Mr Wise himself has also designed and made several badges and insignia.

Wiskemann, S. A.

Belgium, Brussels

40 Rue des Anciens Etangs, Forest. Manufacturing and retail silversmiths. The Wiskemanns were silversmiths in Kassel in the 18th century: firm still possess apprenticeship certificates of 1789 and 1810. By 1805 a branch of the family had already been in Zürich long enough to receive the Freedom. 1852 Otto Leonard Wiskemann was born in Zürich; apprenticed to his father. 1870 went to Paris to study for two years at Christofle factory. There he heard of rival claims of Christofle and Elkington to invention of electroplate (settled in Elkington's favour by Christofle paying Elkingtons two million francs). 1872 Wiskemann started small business in Brussels, introducing the electrolytic process to Belgium. At Vienna 1872 exhibition met and made friends with Gramme, pioneer of the dynamo, and Wiskemann slowly displaced the conventional battery with dynamos for galvanic power for electroplating. An early dynamo dated 1872 is preserved in Wiskemann factory. Awarded honours in international exhibitions in Antwerp (1885, 1889, 1899, 1930), Brussels (1888, 1897, 1900, 1905, 1910, 1935), Liège (1905), Milan (1906), Ghent (1913) and Rio de Janeiro (1923). The British invention of the dynamo by Henry Wilde c. 1860 is now recognized. 1890 firm was moved from its first Brussels premises at Petite Rue and Rue Les Longs Chariots to new factory on site of the convent 'Val des Roses' which was demolished. Later Otto Wiskemann, helped by his sons in the early nineties, built another factory in his native Zürich on Seefeld Street. This exhausted him and he died in Nice 1909. His two sons Otto and Albin took over business, opening new branches in Brussels, Antwerp, Ghent, Liège and Bruges, a factory in Milan 1910, and branches in Nice 1910, and Paris 1930. 1910 another comprehensive new factory built in the Brussels suburb Forest; 1924 firm one of first in the world to manufacture kitchen articles in stainless steel.

1927 Albin Wiskemann founded 'Friendly Society of Goldsmiths and Jewellers Personnel' and became Chairman; this welfare scheme soon spread throughout the country. Since 1923 Albin had also been Brabant director of a family allowances scheme for illness/disability insurance which in 1951 covered over 70,000 families. 1967 firm is managed by M. Soin. Milan and Zürich factories are closed, the Wiskemann family died out after one had for some years joined Wolfers. Seven retail shops in Belgium – Antwerp, Liège, Ghent, Bruges with three in Brussels, also in Paris and Nice. Some four hundred employees of whom two hundred and thirty work in the factory.

Wolfers, Frères

Belgium, Brussels

Court jewelers and silversmiths. In 1812 Edouard Wolfers registered his mark 'W surmounting the cross of Legion of Honour'. Louis Wolfers (1820–92), *maitre et marchand orfèvre*, was prominent after 1850. His mark 'W with a boar's head', is still the mark of the firm. His son was Philippe Wolfers (1858–1929), whose sons were Marcel (sculptor) and Lucien. Lucien's nephew Freddy Wolfers now directs firm. 1956 he began to move factory to Zele where 125 people work in 1967, 125 people remaining in old Horta factory building of 1910 behind Rue d'Aremberg shop. Shop being restored to original Horta design 1967. 1400 sales points for products in Belgium, agency agreement for export of silver to France through Chaumet of Paris. 1967 completed $1 million order for king of Saudi Arabia, weighing 2½ tons of silver, 40 kilos of gold. The highest piece measured 2½ metres.

Wolfers, Philippe 1858–1929

Belgium, Brussels

Sculptor, goldsmith, jeweler. In 1875 at 17 qualified at Beaux Arts Academy where he had studied under Isidore de Rudder; joined family workshops, apprenticed as modeller, chaser, engraver. Louis had no shop, so twice a year Philippe went on sales tours to Belgium, Holland, France, often brought back his drawings for commissions instead of orders for ready-made pieces. Vienna 1873 exhibition had first launched Far Eastern art, including cloisonné enamels, lacquers and engraving with steel, silver and gold mingled. Japanese exhibit in Paris 1889 exhibition achieved fame too: warmly praised by Goncourt and Bing. 1892 first ivory arrived at Antwerp from independent Congo, made available free to sculptors by King Leopold II with view to stimulating new designs. 1894 Philippe Wolfers, one of them, participated in Antwerp International exhibition and was well reviewed in the *Journal de Mons*. Also in 1894 René Lalique first showed his signed personal jewels in the Paris salon. Philippe first worked in neo-rococo style, but by 1895 had turned to *art nouveau* floral and insect motifs in his goldsmith's work. Turned attention to jewelry c. 1897 showed at Brussels World Fair, settings being designed by Van de Velde, Hankar, Serrurier-Bovy, Horta and others. 1897 became art director of family firm, it is therefore impossible to distinguish his work from that of firm, all of which he now concentrated in the *art nouveau* style; 1899 built *art nouveau* style house at La Hulpe, architect of which was Hankar. 1900 style had become more symmetrical and abstract. Represented Turin Exhibition 1902. Jewels all marked at the back 'P.W. exemplaire unique'. Having achieved international fame, 83 different magazines, all over the world, from 1893 to 1908 reproduced his works; in England, 1898, 1899, 1900–1, 1902 *Studio*; 1900–2 *Artist*; 1899, 1901 *Magazine of Art*. Most fertile period followed till end of *art nouveau* in Belgium about 1905. He showed at Munich Sezession in 1898 and 1899, surprisingly not in Paris 1900 World Fair though he showed in the Aublanc gallery Paris. After participating in Turin 1902 International Decorative Art Exhibition, concentrated less on jewels, more on sculpture and silver in styles successively of Louis XV, Louis

XVI and Empire. Last jewel signed 'P.W. exemplaire unique' made in 1904. By 1908 had abandoned jewels altogether for exhibitions, showing just as sculptor; his friend René Lalique did same, setting up his glass works at Combes-la-Ville in same year. 1910 on opening of new premises of firm Wolfers Frères (including their workshops, one of the masterpieces of architect Victor Horta), Philippe Wolfers created large jewels of geometrical-lace design. By 1923 at same time as his neoclassical designs, he was adopting the 'Gioconda-Art Décoratif' style. Such work created sensation at Paris Exhibition 1925. In one room silver, chandeliers, china, table-cloth, furniture, carpet and sculpture were all by him. See *Philippe Wolfers* by his son Marcel Wolfers; Belgian Ministry of Education, 1965. Plates 123 301 and endpaper, pages 9 135

Württembergische Metallwarenfabrik (WMF)
See pages 9 74 142, plates 77–86

Zeitner, Herbert 1900–
Germany
Sculptor and artist-craftsman. Studied 1914–22 Staatl. Zeichenakademie Hanau. Own workshop 1925 Berlin; 1945 Lüneburg. Taught production over 35 years, many exhibitions and distinctions. Silver medal Milan Triennale. Made jewelry since 1919.

Acknowledgements

Many people have helped me with the pictures; first there are the names in the preface; then Kenneth Snowman of Wartski, to whose devoted interest the name of Fabergé owes so much; and then all the designers, and the eminent museums who cherish their pieces. Dag Widman of the National Museum, Stockholm, and Alf Bøe of Oslo both take a particular interest in modern work, and Erik Lassen of Copenhagen has, of course, led the Danish crafts for years. To all of them I record my warm thanks.

Thames and Hudson kindly allowed me to use two pictures from Cassou's *Sources of modern art*. Ward, Lock let me use a passage from James Laver's *Victoriana*. George Rainbird and Michael Joseph released for me part of an article I had written for *Antiques International 1966*.

Silver is a difficult photographic problem, and my biggest debt of gratitude is to the excellent photographers who have turned the problem into a pleasure. The owners of the pieces often took the pictures. Most of the Goldsmiths' Hall, London collection is here represented by Peter Parkinson's photographs. I gladly acknowledge other splendid contributions:

Barbro Littmarck, Stockholm 33, 40, 41, 42; Hans Malmberg, Stockholm 56; Studio Granath, Stockholm 60; Olle, Eskilstuna 59; Hans Hammerskjöld, Stockholm 100, 111; Henk Snoek, London 122; Dagmar Korn, Düsseldorf 127, 351, 353; Gerd Knobloch, Düsseldorf 128–30, Christine Ottewill, Hatfield 131–3; Sundahl, Stockholm 137, 167, 169, 170–182, 184–7; Pietinen, Helsinki 138; Dennis Hooker, London 146, 148, 149; Rolf Hintze, Stockholm 183; James Mortimer, London 243; Teigens, Oslo 246, 298, 303; Architectural Review, London, and Eric de Maré 255, 256; Inge Kitlitschka-Strempel, Klosterneuburg 257, 258; 260; O. Vaering, Oslo 304–306; Jean Collas, Paris 328; Jean-Michel Kollar, Paris 334; Matti A. Pitkänen, Helsinki 335–6; Herta Gebhart, Münster 346; Kunstgewerbeschule, Zürich, 356; Pietinen, Helsinki 360; Wendt, Helsinki 361–2; Studio 51, London 413; Douglas Stevens, Leamington Spa 418; Raymond Wilson, Beaconsfield 417; Council of Industrial Design, London 424; Ounamo, Helsinki 439, 441.

GRAHAM HUGHES

Bibliography

Large original books

Bradbury, Frederick. *History of Old Sheffield Plate.* Macmillan, London 1912. Splendid account of 'origin, growth and decay of the industry' and of silver and Britannia metal.

Casson, J., Langui, E. and Pevsner, N. *The Sources of Modern Art.* Thames and Hudson, London, Munich 1961.

Cremona, Italo. *Il Tempo dell'Art Nouveau.* Valecchi, Florence 1964. Comprehensive but disappointingly little Italian material.

Hughes, Graham. *Modern Jewelry.* Studio Vista, Crown, London, New York 1967; *Erlesener Schmuck.* Otto Maier, Ravensburg 1965.

Rheims, M. *L'art 1900.* Schroll, Vienna, Paris 1965.

Schmutzler, R. *Art Nouveau.* Thames and Hudson, London, Stuttgart 1962.

Small books

Aloi, Roberto. *Esempi di decorazione moderna di tutto il mondo: gioelli, sbalzi, argenti.* Hoepli, Milan 1954. Useful but uneven large pictorial booklet.

Amaya, Mario. *Art Nouveau.* Studio Vista, Dutton, London, New York 1966.

Baden-Württemberg, neue Goldschmiedekunst in. Dr Max Löffler, Kohlhammer, Stuttgart. A conservative pictorial selection.

Burch-Korrodi, Meinrad. *Gold- und Silberarbeiten aus der Werkstatt Meinrad Burch-Korrodi.* NZN, Zürich 1954.

David-Andersen 1876–1951. Oslo. The firm's memorial book.

De Fusco, Renato. *Il Floreale a Napoli.* Edizioni Scientifiche Italiane, Naples 1959.

Du magazine, William Morris issue, September 1965. Atlantis, Zürich. The best German short survey in existence.

Gestaltendes Handwerk. Deutsche Handwerks-Institut, Bonn 1963. A convincing pictorial survey of German craftwork.

Glenister, S. J. and Larkman, Brian. *Contemporary design in metalwork.* Murray, London 1963. A short introduction to British work.

Haycraft, John. *Finnish jewellery and silverware: an introduction to contemporary work and design.* Helsinki 1960.

Himsworth, J. B. *The story of cutlery from flint to stainless steel.* Ernest Benn, London 1953. Mostly covers early periods.

Hiort, Esbjørn. *Modern Danish Silver.* New York, London, Stuttgart, Teufen and Copenhagen 1954.

Hughes, George Ravensworth. *The Worshipful Company of Goldsmiths as patrons of their craft.* Goldsmith's Hall, London 1965. Invaluable account of the revival of British silversmithing largely under that author's influence.

Hughes, Graham. *Jewelry.* Studio Vista, Dutton, London, New York 1966.

Koczogh, Akos. *Modern Hungarian metalwork.* Budapest 1964.

Leipzig. *Report on the international craft exhibition held in Leipzig city museum 1927.* E. A. Seeman, Leipzig 1928. Illustrations of absorbing interest and distinction.

Lotz, Wilhelm. *Modern German gold and silver.* Picture booklet. Hermann Reckendorf, Berlin 1926.

Møller, Viggo Sten. *Henning Kopel* (Danish, English and German). Carlsberg Foundation, Copenhagen 1965.

Noyes, Pierrepont B. *My Father's House,* 1937. *A Goodly Heritage,* 1958. Rinehart, New York, Toronto. Oneida's early history.

Persson, Sigurd. *Modern Swedish silver.* Lindquist, Stockholm 1951.

Rheims, Maurice. *L'Obet 1900. Arts et Métiers Graphiques.* Paris 1964. Some outstanding creations shown in colour.

Scheidig, Walther. *Crafts of the Weimar Bauhaus.* Studio Vista, Edition Leipzig, London 1967, Leipzig 1966.

Schwartz, Walter. *Georg Jensen* (Danish). Jensen and Wendel, Copenhagen 1958.

Snowman, Kenneth. *The art of Carl Fabergé.* Faber, London 1962.

Stavenow. Åke. *Silversmeden Jacob Ångman 1876–1942.* Nordisk Rotogravyr, Stockholm 1955. Admirable biographical study in Swedish.

Sutherland, C. H. V. *Gold.* Thames and Hudson, London 1959. Fine historical survey.

Taylor, Gerald. *Art in Silver and Gold.* Studio Vista, Dutton, London, New York 1964. *Silver.* Penguin, London 1963. Both mostly antiques.

Wardle, Patricia. *Victorian Silver.* Herbert Jenkins, London 1963.

Catalogues

Brussels
Le Bijou 1900. Hotel Solvay, 1965.

Copenhagen
Fifty years of Danish silver in the Georg Jensen tradition. Jensen, 1954.

Jensen, Georg. Centenary exhibition (1866–1966). Goldsmith's Hall, London. *Jensen Sølvsmedie, Mobilia* magazine. Text in English, French, Danish, German.

Darmstadt
Bott, Gerhard. *Kunsthandwerk um 1900. Jugendstil, art nouveau.* 1965. Admirable large catalogue of museum's outstanding permanent collection.

Frankfurt am Main
Messer Löffel Gabel. Museum für Kunsthandwerk. 1964.

The Hague
Nederlands zilver, 1815–1960. Gemeentemuseum. 1960.

Hanau
Hanau – historisches und neues Kirchengerät. Goldschmiedehaus Hanau. 1965.

London
Arts Council of Great Britain. *Art Nouveau in Britain.* 1965. Goldsmith's Hall.
'Mucha', Grosvenor Gallery, Victoria and Albert Museum, 1963.

British silverwork by contemporary craftsmen. 1951.

Modern British Silver. 1963. Picture booklet.

New gold, silver and jewels commissioned by industry. 1965.

Becker, Friedrich. 1966

Copy or Creation – Victorian Treasures from English Churches. Ed. Shirley Bury. 1967.

The Goldsmith Today. 'Acquisitions of new silver and jewels since 1953 by the Worshipful Company of Goldsmiths.' 1967.

Milan
Art Deco 1920–1930. Galleria Milano, 1965.

Monza
Monza 1925. Pictures of the heavy German crafts exhibit at the international display in Monza, in the year which also saw the huge Paris exhibition. Hermann Reckendorf, Berlin 1926.

New York
Tiffany, Louis Comfort, 1848–1933. Museum of Contemporary Crafts, 1958.

Oslo
Arbok, 1959–1962. Kunstindustrimuseet.

Ostend
Europa 1900. 1967.

Paris
Les Années '25'. Musée des Arts Décoratifs, 1966.

Stuttgart
Internationales Kunsthandwerk 1966. LGA Museum.

Vienna
Wien um 1900 Oesterreichisches Museum für angewandte Kunst, 1964.

Zeist
Europees Handgedreven Zilver. Showroom van de Koninklijke van Kempen & Begeer. Holland 1965. Also published abbreviated in German for Hanover showing.

255

Index

Plate numbers are indicated by italic figures

Abbo, (Jussuf) Hussein *409*
American Craftsmen's Council 72
Aubock, Karl *436*
Ballin, Mogens 11, 93
Bauhaus *325, 329*
Berlage, H. P. *282*
Bernstein, Bernard *345*
Beyschlag *77, 80–1*
Bing, S. 52
Blunt & Wray *323*, 131, 139
Borgila *see* Fleming
Bolek, Hans *264*
Brandt, Marianne *325, 329*
Brandtberg, Oscar *34, 37–9*
Brauchi & Fils *359*
Christiansen, Prof. *283*
Crafts Centre of Great Britain 73–4
Dawson City 68
De Beers Consolidated Mines Ltd *403*, 74, 139
Ditzel, Nanna (b. 1923) 12
Ehrström, Eric (1881–1934) *299, 302*
Eisenberg-Lozano Inc. *103*
Engelhardt, Knud *309*, 11
Ericsson, Henry (1898–1933) *335*
Fenster, Fred *344*
Fisker, Kay *339*
Gallé, Émile 9, 136
Gense, *see* GAB 76
Glover, Robert *427*
Gretsch *64*
Hablic, Wunzel *326*
Haglund, Birger *98*
Herløw, Erik *435*
Hertz, P. *308*
Ionian Bank 38
Irvine, Alan *134–6*
Jacobsen, Arne *437*
Jung, Gunilla *336*
Kalgoorlie 68
Kayser Sohn J. P. *296–7*
Kayserzinn 132
Klondike 68
Knapp, Stefan 141

Knox, Alexander *248*
Konstfackskolan *97, 98, 100, 103, 111*, 31, 55, 97
Konstflitföreningen 72
Laver, James 130
Leven, Hugo *296–7*
Liljedahl, Bengt *100*
Manzù, Giacomo (b. 1908) *419–421*
Mappin, David 77
Mau, M. *290*
Miller, F. A. *343*
Moore, John C., Edward C. 49, 52
Morris, William 7, 9; 10, 52, 53, 72, 93, 129, 133, 134
Müller, Albin *295*
Neuzeughammer *434, 436*
Petrini, Alexander *97*
Pfefferman, Kurt *348*
Pogacnic, Theodor *266*
Ponti, Gio *327, 433*
Prouvé, Victor (1858–1943) *93*, 9, 136
Prytz, Torolf, Jacob *see* Tostrup 134
Roberts & Belk *see* C. J. Vander 75
Ruskin, John 52
Schools of art, history of 55
Schwäbisch Gmünd *101*
Sirnes, Johan *305–6*
Skoogfors, Olof *347*
Slöjdföreningen 72
Sutherland, Graham *413*
Templier, Raymond (b. 1891) *328*
Universal Steel Company *435*
Vever, Henri 136
Vignando, Luigi *359*
Wahlman, L. I. *50*
Wheeler, Peter *383*
Wiener, Werkstätte 134
Wilm, Johann, Michael (b. Munich 1885) 138
Wimmer, Eduard *267*
Worshipful Company of Goldsmiths *134, 136*, 25, 47, 58, 74, 99, 139, 140
Wostenholm, George 75
Wynne, David *422*
Zehrer, Max *352*

Nassau Library System

Roosevelt Field
Garden City, N. Y.

A cooperative system of Libraries created to foster
quality and develop increased library service to the
residents of local Communities in Nassau County.

70

PRINTED IN U.S.A.